Lecture Notes in Mathematics

Edited by A. Dold and B. Eckmann

849

Péter Major

Multiple Wiener-Itô Integrals

With Applications to Limit Theorems

Springer-Verlag
Berlin Heidelberg New York 1981

Author

Péter Major
Mathematical Institute
The Hungarian Academy of Sciences
Reáltanoda u. 13–15
1053 Budapest
Hungaria

AMS Subject Classifications (1980): 60G15, 60G20, 60H05

ISBN 3-540-10575-1 Springer-Verlag Berlin Heidelberg New York
ISBN 0-387-10575-1 Springer-Verlag New York Heidelberg Berlin

Printing and binding: Beltz Offsetdruck, Hemsbach/Bergstr.
2141/3140-543210

TABLE OF CONTENTS

INTRODUCTION

One of the most important problems in probability theory is the investigation of the limit distribution of partial sums of appropriately normalized random variables. The case where the random variables are independent is fairly well understood. Many results are known also in the case where independence is replaced by an appropriate mixing condition or some other "almost independence" property. Much less is known about the limit behaviour of partial sums of really dependent random variables. On the other hand, this case is becoming more and more important, not only in probability theory, but also in some applicatons to statistical physics.

The problem about the asymptotic behaviour of partial sums of dependent random variables leads to the investigation of some very complicated transformations of probability measures. The classical methods of probability theory do not seem to work for this problem. On the other hand, although we are still very far from a satisfactory solution of this problem, we can already present some nontrivial results.

The so-called multiple Wiener-Itô integrals have proved to be a very useful tool in the investigation of this problem. The proofs of almost all rigorous results in this field are closely related to this technique. The notion of multiple Wiener-Itô integrals was worked out for the investigation of nonlinear functionals over Gaussian fields. It is closely related to the so-called Wick polynomials which can be considered as the multidimensional generalization of Hermite polynomials. The notions of Wick polynomials and multiple Wiener-Itô integrals were worked out at about the same time and independently of each other. Actually, we discuss a modified version of multiple Wiener-Itô integrals in greatest detail. The technical changes needed in the definition of these modified integrals are not essential. On the other hand, these modified integrals are more appropriate for certain investigations, since they enable us to describe the action of shift transformations and to apply some sort of random Fourier analysis. There is also some connection between multiple Wiener-Itô integrals and the classical stochastic Itô integrals. The main difference between them is that in the first case deterministic functions are integrated, in the second case so-called non-anticipating functionals. One consequence of this difference is that no technical difficulty arises when we want to define multiple Wiener-Itô integrals in the multidimensional time case. On the other hand, a large class of nonlinear functional over Gaussian fields can be represented by means of multiple Wiener-Itô integrals.

In this work we are mainly interested in limit problems for sums of dependent random variables. It is useful to consider this problem together with its continuous time version. The natural formulation of the continuous time version of this problem can be given by means of generalized fields. Consequently we also have to discuss some questions about generalized fields.

I have not tried to formulate all the results in the most general form. My main goal was to work out the most important techniques needed in the investigation of such problems. This is the reason why the greatest part of this work deals with multiple Wiener-Itô integrals. I have tried to give a self-contained exposition of this subject and also to explain the motivation behind the results.

I had the opportunity to participate in the Dobrushin - Sinai seminar in Moscow. What I learned there was very useful also for the preparation of this Lecture Note. Therefore I would like to thank the members of this seminar for what I could learn from them, especially P.M.Blecher, R.L. Dobrushin and Ja.G.Sinai.

1) On a limit problem

We begin with the formulation of a problem, which is important both for probability theory and statistical physics. The multiple Wiener-Itô integral proved to be a very useful tool at the investigation of this problem.

Let ξ_n, $n \in Z_\nu$, where Z_ν denotes the ν-dimensional integer lattice, be a discrete (strictly) stationary random field. We recall that a set of random variables ξ_n, $n \in Z_\nu$, is called a (discrete) random field. It is called (strictly) stationary if $(\xi_{n_1}, \dots, \xi_{n_k}) \stackrel{\Delta}{=} (\xi_{n_1+m}, \dots, \xi_{n_k+m})$ for all $k=1,2,\dots$ and $n_1, \dots, n_k, m \in Z_\nu$, where $\stackrel{\Delta}{=}$ denotes equality in distribution. For all $N=1,2,\dots$ we define the new fields

$$(1.1) \qquad Z_n^N = A_N^{-1} \sum_{j \in B_n^N} \xi_j \,, \qquad N=1,2,\dots \qquad n \in Z_\nu$$

where

$$B_n^N = \{j \mid j \in Z_\nu\,,\ n^{(i)}N \le j^{(i)} < (n^{(i)}+1)N, i=1,2,\dots,\nu\}$$

(the superscript i denotes the i-th coordinate of a vector), and A_N is an appropriate norming constant.

We are interested in the following questions:

When do the finite dimensional distributions of the fields Z_n^N converge to the finite dimensional distributions of a field Z_n^*? Which fields Z_n^* can appear as limits?

During the investigation of the above questions one also has to solve the following problem: Which fields ξ_n satisfy the relation

$$(1.2) \qquad (\xi_{n_1}, \ldots, \xi_{n_k}) \stackrel{\Delta}{=} (Z^N_{n_1}, \ldots, Z^N_{n_k})$$

for all $N=1,2,\ldots$ and $n_1, \ldots, n_k \in Z_\nu$. If the field ξ_n satisfies relation (1.2) with $A_N = N^\alpha$ then ξ_n (or its distribution) is called a self-similar field with self-similarity parameter α. We are interested in the case $A_N = N^\alpha$ because, under some slight restrictions, for fields ξ_n satisfying (1.2) A_N must be chosen in this way. A central problem both in statistical physics and in probability theory is the descriprion of self-similar fields. We are interested in self-similar fields whose random variables have a finite second moment. This excludes the fields consisting of i.i.d. random variables with a non-Gaussian stable law.

The Gaussian self-similar fields and their Gaussian range of attraction are fairly well known. Much less is known about the non-Gaussian case. The problem is hard, because the transformations of measures over R^{Z_ν} induced by formula (1.1) have a very complicated structure. We shall define the so-called subordinated fields below. (More precisely the fields subordinated to a stationary Gaussian field). In case of subordinated fields the Wiener-Itô integral is a very useful tool for investigating the transformation defined in (1.1). In particular, it enables one to construct non-Gaussian self-similar fields and to prove non-trivial limit theorems. All known rigorous constructions of self-similar fields are closely related to this technique.

Let X_n, $n \in Z_\nu$, be a stationary Gaussian field. We define the shift transformations T_m, $m \in Z_\nu$, over this field by the formula $T_m X_n = X_{m+n}$ for all $n, m \in Z_\nu$. Let H denote the real Hilbert space consisting of the square integrable random variables measurable with respect to the σ-algebra $\mathcal{B} = \mathcal{B}(X_n, n \in Z_\nu)$. The scalar product in H is defined as $(\xi,\eta) = E\xi\eta$, $\xi,\eta \in H$. The shift transformations T_m, $m \in Z_\nu$, can be extended to a group of unitary shift transformations over H in a natural way. Namely, if $\xi = f(X_{n_1}, \dots, X_{n_k})$ then we define $T_m\xi = f(X_{n_1+m}, \dots, X_{n_k+m})$. It can be seen that $\|\xi\| = \|T_m\xi\|$, and the above considered random variables ξ are dense in H. Hence T_m can be extended to the whole space H by L_2-continuity. Now we introduce the following

Definition

Given a discrete stationary Gaussian field X_n, $n \in Z_\nu$, we define the Hilbert space H and the shift transformations T_m, $m \in Z_\nu$, over H as before. The discrete stationary fields ξ_n is called a random field subordinated to X_n if $\xi_n \in H$, and $T_n\xi_m = \xi_{n+m}$ for all $n,m \in Z_\nu$.

We remark that ξ_o determines the subordinated field ξ_n completely, since $\xi_n = T_n \xi_o$. Later we give a more adequate description of subordinated fields by means of Wiener-Itô integrals. Before working out the details we formulate the continuous time version of the above notions and problems. In the continuous time case it is more natural to consider generalized fields.

Let $S=S_\nu$ be the Schwartz space of (real valued) rapidly decreasing functions over R^ν (See e.g. [14] for the definition of S_ν). We shall omit the subscript ν if it leads to no ambiguity. We say that the set of random variables $X(\varphi)$, $\varphi \in S$, is a generalized field over S if:

a) $X(a_1\varphi_1+a_2\varphi_2) = a_1X(\varphi_1) + a_2X(\varphi_2)$ for all real numbers a_1, a_2 and $\varphi_1, \varphi_2 \in S$.

b) $X_n(\varphi) \Rightarrow X(\varphi)$ stochastically if $\varphi_n \to \varphi$ in the topology of S.

The field $X = \{X(\varphi),\ \varphi \in S\}$ is stationary if $X(\varphi) \triangleq X(T_t\varphi)$ for all $\varphi \in S$ and $t \in R^\nu$, where $T_t\varphi(x) = \varphi(x+t)$, it is Gaussian if $X(\varphi)$ is a Gaussian random variable for all $\varphi \in S$. $X_n \xrightarrow{\mathcal{D}} X_o$ as $n \to \infty$ if $X_n(\varphi) \xrightarrow{\mathcal{D}} X_o(\varphi)$ for all $\varphi \in S$, where $\xrightarrow{\mathcal{D}}$ denotes convergence in distribution. Given a stationary generalized field X and a function $A(t) > 0$, $t > 0$, we define the fields X_t^A for all $t > 0$ by the formula

$$(1.3)\qquad X_t^A(\varphi) = X(\varphi_t^A),\quad \varphi \in S,\quad \text{where } \varphi_t^A(x) = A(t)^{-1}\varphi\left(\frac{x}{t}\right).$$

We are interested in the following

Question

When does a field X^* exist such that $X_t^A \xrightarrow{\mathcal{D}} X^*$ as $t \to \infty$ (or $t \to 0$)?

We introduce the following

Definition

The stationary generalized field X is self-similar with self-similarity parameter α if $X_t^A(\varphi) \stackrel{\Delta}{=} X(\varphi)$ for all $\varphi \in S$ and $t > 0$ with $A(t) = t^\alpha$.

To answer the above question one should first describe the generalized self-similar fields.

We try to explain the motivation behind the above definitions. Given an ordinary field $X(t)$, $t \in R^\nu$, and a topological space E consisting of functions over R^ν one can define the random variables $X(f) = \int f(t)X(t)dt$, $f \in E$. Some difficulty may arise when defining this integral, but it can be overcome in all interesting cases. If the space E is rich enough, and this is the case if $E=S$, then the integrals $X(f)$, $f \in E$, determine the process $X(t)$. The set of the random variables $X(f)$, $f \in S$, is a generalized field in all nice cases. On the other hand, there are generalized fields which cannot be obtained by integrating an ordinary field. In particular, the generalized self-similar fields we shall construct later cannot be interpreted through ordinary fields. The above definitions of various properties of generalized fields are rather natural, considering what these definitions mean for generalized fields obtained by integrating ordinary fields. The investigation of generalized fields is simpler than that of ordinary discrete fields, because in the continuous case there is more symmetry available. Moreover, even when discrete fields are investigated, integration with respect to a random measure defined through a generalized field is often useful.

We finish this section by defining the generalized subordinated fields. Let X be a generalized stationary Gaussian field. The formula $T_t(X(\varphi)) = X(T_t\varphi)$ defines the shift transformations T_t for all $t \in R^\nu$. Let H denote the real Hilbert space consisting of the $\mathcal{B}=\mathcal{B}(X(\varphi), \varphi \in S)$ measurable random variables with finite second moment. The shift transformations T_t can be extended to be a group of unitary transformations T_t over H, similarly to the discrete case.

Definition

Given a generalized stationary Gaussian field X, we define the Hilbert space H and the shift transformations over H as above. The generalized stationary field ξ is subordinated to X if $\xi(\varphi) \in H$, $T_t\xi(\varphi) = \xi(T_t\varphi)$ for all $\varphi \in S$ and $t \in R^\nu$, and $E[X(\varphi_n) - X(\varphi)]^2 \to 0$ if $\varphi_n \to \varphi$ in the topology of S.

2. Wick polynomials

In this section we consider the so-called Wick polynomials, a multi-dimensional generalization of Hermite polynomials. They are closely related to multiple Wiener-Itô integrals.

Let X_t, $t\in T$, be a set of jointly Gaussian random variables indexed by a parameter set T. Let $EX_t = 0$ for all $t\in T$. We define the real Hilbert spaces H_1 and H in the following way: A square integrable random variable is in H if and only if it is measurable with respect to the σ-algebra $\mathcal{B}=\mathcal{B}(X_t, t\in T)$, and the scalar product in H is defined as $(\xi,\eta) = E\xi\eta$, $\xi,\eta\in H$. $H_1\subset H$ is the subspace generated by the finite linear combinations $\sum c_j X_{t_j}$, $t_j\in T$. We consider only such sets X_t, $t\in T$, for which H_1 is separable. $\{X_t, t\in T\}$ can be otherwise arbitrary, but the most interesting case is when $T=S_\nu$ or Z_ν, and X_t, $t\in T$, is a stationary Gaussian field.

Let $Y_1, Y_2, \dots$ be an orthonormal basis in H_1. Then the uncorrelated random variables $Y_1, Y_2, \dots$ are independent, because they are Gaussian. Moreover $\mathcal{B}(Y_1, Y_2, \dots) = \mathcal{B}(X_t, t\in T)$. Let $H_n(x)$ denote the n-th Hermite polynomial with leading coefficient 1, i.e. let $H_n(x)=(-1)^n \exp(\frac{x^2}{2}) \frac{d^n}{dx^n} (\exp(-\frac{x^2}{2}))$. We recall the following results from analysis and measure theory.

<u>Theorem 2A</u>

<u>The Hermite polynomials $H_n(x)$, $n=0,1,2,\dots$ form a complete orthogonal system in</u> $L_2(R,\mathcal{B}, \frac{1}{\sqrt{2\pi}} \exp(-\frac{x^2}{2})\, dx)$. (Here $\mathcal{B}$ denotes the Borel σ-algebra on the real line).

Let $(X_j, \mathcal{X}_j, \mu_j)$, $j=1,2,\dots$ be countably many copies of a probability space $(X,\mathcal{X},\mu)$. (We denote the points of X_j by x_j.) Let $(X^\infty, \mathcal{X}^\infty, \mu^\infty) = \prod_{j=1}^\infty (X_j, \mathcal{X}_j, \mu_j)$.

Theorem 2B

Let $\varphi_o, \varphi_1, \ldots; \varphi_o(x) \equiv 1$ be a complete orthonormal system in $L_2(X,\chi,\mu)$. Then the functions $\prod_{j=1}^{\infty} \varphi_{k_j}(x_j)$, where only finitely many indices k_j differ from zero, form a complete orthonormal basis in $(X^\infty, \chi^\infty, \mu^\infty)$.

Theorem 2C

Let (X,A) be a measurable space, $Y_1, Y_2, \ldots$ be A-measurable functions such that $B(Y_1,Y_2,\ldots)=A$. If ξ is an A-measurable function then there exists an (R^∞, B^∞) - measurable function f such that $\xi = f(Y_1,Y_2,\ldots)$.

Theorem 2A, 2B and 2C have the following important consequence:

Theorem 2.1

Let $Y_1, Y_2, \ldots$ be an orthonormal basis in H_1. Then the set of all possible finite products $H_{j_1}(Y_{\ell_1}) \ldots H_{j_k}(Y_{\ell_k})$ is a complete orthogonal system in H.

Proof of Theorem 2.1

By Theorems 2A and 2B, the set of all possible products $\prod_{j=1}^{\infty} H_{k_j}(x_j)$, where only finitely many indices k_j differ from 0, is a complete orthogonal system in $L_2(R^\infty, B^\infty, \prod_{j=1}^{\infty} \frac{\exp(-\frac{x_j^2}{2})}{\sqrt{2\pi}} dx_j)$. Since $B(X_t, t \in T) = B(Y_1, Y_2, \ldots)$, Theorem 2C implies that the mapping $f(x_1, x_2, \ldots) \to f(Y_1, Y_2, \ldots)$ is a unitary transformation from

$L_2(R^\infty, B^\infty, \prod_{j=1}^{\infty} \frac{dx_j}{\sqrt{2\pi}} \exp(-\frac{x_j^2}{2}))$ to H. Since the image of a complete orthogonal system under a unitary transformation is again a complete orthogonal system, Theorem 2.1 is proved.

Let $H_{\leq n} \subset H$, $n=1,2,\ldots$ denote the Hilbert space which is the closure of the linear space consisting of the elements $P_n(X_{t_1},\ldots,X_{t_n})$ where P_n runs through all polynomials of order less than or equal to n, and the integer m and the indices $t_1,\ldots,t_m \in T$ are arbitrary. Let $H_o=H_{\leq o} \subset H$ consist of the constant functions, and let $H_n=H_{\leq n} \ominus H_{\leq n-1}$ where $\ominus$ denotes orthogonal completition. It is clear that the two definitions of H_1 given in this section coincide. If $\xi_1,\ldots,\xi_m \in H_1$, and $P_n(x_1,\ldots,x_m)$ is a polynomial of order n, then $P_n(\xi_1,\ldots,\xi_m) \in H_{\leq n}$. Hence Theorem 2.1 implies that

$$H=H_o + H_1 + H_2 + \ldots \tag{2.1}$$

where $+$ denotes direct sum. Now we introduce the following

Definition

Given a polynomial $P(x_1,\ldots,x_m)$ of order n and the random variables $\xi_1,\ldots,\xi_m \in H_1$, the Wick polynomial $:P(\xi_1,\ldots,\xi_m):$ is the orthogonal projection of $P(\xi_1,\ldots,\xi_m)$ to H_n.

It is clear that Wick polynomials of different order are orthogonal. Given some $\xi_1,\ldots,\xi_m \in H_1$, define $H_{\leq n}(\xi_1,\ldots,\xi_m) \subseteq H_{\leq n}$, $n=1,2,\ldots$ as the set of polynomials

$P(\xi_1,\dots,\xi_m)$ of $\xi_1,\dots,\xi_m$ with order less than or equal to n. Let $H_{\le o}(\xi_1,\dots,\xi_m) = H_o(\xi_1,\dots,\xi_m) = H_o$ and $H_n(\xi_1,\dots,\xi_m) = H_{\le n}(\xi_1,\dots,\xi_m) \ominus H_{\le(n-1)}(\xi_1,\dots,\xi_m)$. Now we formulate the following

Proposition 2.2

Let $P(x_1,\dots,x_m)$ be a polynomial of order n. Then $:P(\xi_1,\dots,\xi_m):$ equals the orthogonal projection of $P(\xi_1,\dots,\xi_m)$ to $H_n(\xi_1,\dots,\xi_m)$.

Proof of Proposition 2.2

Let $:\bar{P}(\xi_1,\dots,\xi_m):$ denote the projection of $P(\xi_1,\dots,\xi_m)$ to $H_n(\xi_1,\dots,\xi_m)$. Obviously

$$P(\xi_1,\dots,\xi_m) - :\bar{P}(\xi_1,\dots,\xi_m): \in H_{\le n-1}(\xi_1,\dots,\xi_m) \subseteq H_{\le n-1} .$$

Hence in order to prove Proposition 2.2 it is enough to show that for all $\eta \in H_{\le n-1}$

(2.2) $$E:\bar{P}(\xi_1,\dots,\xi_m):\eta = 0$$

Let $\varepsilon_1,\varepsilon_2,\dots$ be an orthogonal system in H_1, also orthogonal to $\xi_1,\dots,\xi_m$, and such that $\xi_1,\dots,\xi_m,\varepsilon_1,\varepsilon_2,\dots$ form a basis in H_1. If $\eta = \prod_{i=1}^{m} \xi_i^{\ell_i} \cdot \prod_{j=1}^{\infty} \varepsilon_j^{k_j}$, $\sum \ell_i + \sum k_j \le n-1$ then (2.2) holds true because of the independence of the random variables ξ

and ε. Since the linear combinations of such η are dense in $H_{\leq n-1}$, formula (2.2) and Proposition 2.2 are proved.

Corollary 2.3

Let $\xi_1,\ldots,\xi_m$ be an orthonormal system in H_1. Let $P(x_1,\ldots,x_m) = \sum c_{j_1,\ldots,j_m} x_1^{j_1} \ldots x_m^{j_m}$ be a homogeneous polynomial. Then

$$:P(\xi_1,\ldots,\xi_m): = \sum c_{j_1,\ldots,j_m} H_{j_1}(\xi_1)\ldots H_{j_m}(\xi_m)$$

In particular

$$:\xi^n: = H_n(\xi) \quad \text{if} \quad \xi\in H_1 \quad \text{and} \quad E\xi^2=1 .$$

Proof of Corollary 2.3

Let the order of the polynomial P be n. Then

$$P(\xi_1,\ldots,\xi_m) - \sum c_{j_1,\ldots,j_m} H_{j_1}(\xi_1)\ldots H_{j_m}(\xi_m)\in H_{\leq n-1}(\xi_1,\ldots,\xi_m)$$

since $P(x_1,\ldots,x_m) - \sum c_{j_1,\ldots,j_m} H_{j_1}(x_1)\ldots H_{j_m}(x_m)$ is a polynomial of order less than n. Let $\eta=\xi_1^{\ell_1}\ldots\xi_m^{\ell_m}$, $\sum_{i=1}^{m} \ell_i\leq n-1$. Then

$$E\eta H_{j_1}(\xi_1)\ldots H_{j_m}(\xi_m) = \prod_{i=1}^{m} \xi_i^{\ell_i} H_{j_i}(\xi_i) = 0$$

since $\ell_i < j_i$ for at least one index i. Therefore

$$E\eta \sum c_{j_1,\ldots,j_m} H_{j_1}(\xi_1)\ldots H_{j_m}(\xi_m) = 0 .$$

Since every element of $H_{\leq n-1}(\xi_1,\ldots,\xi_m)$ can be written as the linear combination of such elements η, the above relations together with Proposition 2.2 imply Corollary 2.3.

Corollary 2.4

Let $\xi_1,\xi_2,\ldots$ be an orthonormal basis in H_1. Then the random variables $H_{j_1}(\xi_{l_1})\ldots H_{j_k}(\xi_{\ell_k})$, $j_1+\ldots+j_k = n$ form a complete orthogonal system in H_n.

The arguments of this section exploited heavily some properties of Gaussian random variables. Namely, that the linear combination of Gaussian variables is again Gaussian, and in Gaussian case orthogonality implies independence. This means, in particular, that the rotation of a standard normal vector leaves its distribution invariant. We finish this section with an observation based on these facts, which may illuminate the content of formula (2.1) from another point of view.

Let U be a unitary transformation over H_1. It can be extended to a unitary transformation $\mathcal{U}$ over $\mathcal{H}$ in a natural way. Fix an orthonormal basis $\xi_1,\xi_2,\ldots$ in H_1. and define $\mathcal{U}1=1$, $\mathcal{U}\xi_{j_1}^{\ell_1}\ldots\xi_{j_k}^{\ell_k} = (U\xi_{j_1})^{\ell_1}\ldots(U\xi_{j_k})^{\ell_k}$.

This transformation can be extended to a linear transformation $\mathcal{U}$ over $\mathcal{H}$ in a unique way. The transformation $\mathcal{U}$ is norm preserving since the joint distributions of $(\xi_1,\xi_2,\ldots)$ and $(U\xi_1,U\xi_2,\ldots)$ coincide. Moreover it is unitary since $U\xi_1,U\xi_2,\ldots$ is an orthonormal basis in H_1. It is not difficult to see that if $P(x_1,\dots,x_m)$ is an arbitrary polynomial and $\eta_1,\eta_2,\dots,\eta_m \in H_1$, then $\mathcal{U}P(\eta_1,\dots,\eta_m) = P(U\eta_1,\dots,U\eta_m)$. This relation means in particular that the transformation $\mathcal{U}$ does not depend on the choice of the basis in H_1. If the transformations $\mathcal{U}_1$ and $\mathcal{U}_2$ correspond to U_1 and U_2 then the transformation $\mathcal{U}_1\mathcal{U}_2$ corresponds to U_1U_2. The subspaces $H_{\le n}$ and therefore the subspaces H_n remain invariant under these transformations $\mathcal{U}$. The shift transformations of a stationary Gaussian field, and their extensions to H are the most interesting examples for such unitary transformations U and $\mathcal{U}$. In the terminology of group representations the above facts can be formulated in the following way: The mapping $U \to \mathcal{U}$ is a group representation of $U(H_1)$ over H, where $U(H_1)$ denotes the group of unitary transformations over H_1. Formula (2.1) gives a decomposition of H into orthogonal invariant subspaces of this representation.

3. Random spectral measures

Some standard theorems of probability theory state that the correlation function of a stationary field can be expressed as the Fourier transform of a so-called spectral measure. In this section we construct a random measure with the help of these results, and we express the random field itself as the Fourier transform of this random measure in some sense. We restrict ourselves to the Gaussian case, although most of the results in this section are valid for arbitrary stationary field with finite second moment if independence is substituted by orthogonality. In the next section we define the multiple Wiener-Itô integrals with respect to this random measure. In the definition of multiple stochastic integrals the Gaussian property will be heavily exploited. First we recall two results about the spectral representation of the covariance function. Given a stationary Gaussian field X_n, $n \in Z_\nu$, or $X(\varphi)$, $\varphi \in S$, we shall assume throughout the paper the $EX_n = 0$, $EX_n^2 = 1$ in the discrete and $EX(\varphi) = 0$ in the generalized case.

Theorem 3A (Bochner)

Let X_n, $n \in Z_\nu$, be a discrete (Gaussian) stationary field. There exists a unique probability measure G on $[-\Pi, \Pi)^\nu$ such that the correlation function $r(n) = EX_0 X_n$, $n \in Z_\nu$, can be written in the form

$$r(n) = \int \exp[i(n,x)] G(dx) ,$$

where $(.,.)$ denotes scalar product. Further, $G(A) = G(-A)$ for all $A \subset [-\Pi, \Pi)^\nu$.

(We identify $[-\Pi, \Pi)^\nu$ with the torus $R^\nu / 2\Pi Z_\nu$. Thus e.g. $-(-\Pi, \ldots, -\Pi) = (-\Pi, \ldots, -\Pi)$.)

Theorem 3B (Bochner-Schwartz)

Let $X(\varphi)$, $\varphi \in S$, be a generalized (Gaussian) stationary field over $S=S_\nu$. There exists a unique σ-finite measure G on R^ν such that

$$EX(\varphi)X(\Psi) = \int \tilde{\varphi}(x)\overline{\tilde{\Psi}(x)}G(dx) \quad \text{for all} \quad \varphi,\Psi\in S,$$

where ~ denotes Fourier transform. The measure G has the properties $G(A) = G(-A)$ for all $A\in B^\nu$, and

$$\int (1+|x|)^{-r}\, G(dx) < \infty$$

with an appropiate $r>0$.

The measure G appearing in Theorems A and B is called the spectral measure of the stationary random field. A measure G with the same properties as the measure G in Theorem 3A or 3B will also be called a spectral measure. This terminology is justified since there exists a random field with spectral measure G for all such G.

Let us now consider a stationary Gaussian random field (discrete or generalized one) with spectral measure G. We shall denote the space $L_2([-\Pi,\Pi)^\nu, B^\nu, G)$ or $L_2(R^\nu, B^\nu, G)$ simply by L_G^2. Let H_1 be the real Hilbert space defined by means of this stationary field in the same way as it was done in Section 2. Let H_1^c denote its complexification, i.e. the elements of H_1^c are of the form $\varphi+i\psi$, $\varphi,\psi\in H_1$, and $(\varphi_1+i\psi_1, \varphi_2+i\psi_2) = (\varphi_1,\varphi_2) + (\psi_1,\psi_2) + i[(\psi_1,\varphi_2)-(\varphi_1,\psi_2)]$. We are going to construct a unitary transformation I from L_G^2 to H_1^c.

We shall define the random spectral measure via this transformation.

Let S^c denote the Schwartz space of rapidly decreasing complex valued functions with the usual topology. (The elements of S^c are of the form $\varphi+i\psi$, $\varphi,\psi\in S$.)

We make the following observation. The finite linear combinations $\sum c_n \exp[i,(n,x)]$ are dense in L_G^2 in the discrete case, the functions $\varphi\in S^c$ are dense in L_G^2 in the generalized case. In the discrete case this follows immediately from the Weierstrass approximation theorem, which states that all continuous functions over $[-\Pi,\Pi)^\nu$ can be uniformly approximated by trigonometrical polynomials. In the generalized case let us first observe that the continuous functions with compact support are dense in L_G^2. We claim that also the functions $\varphi\in\mathcal{D}$ are dense in L_G^2, where $\mathcal{D}$ denotes the class of (complex valued) infinitely many times differentiable functions with compact support. Indeed, if $\varphi\in\mathcal{D}$, φ is real valued, $\varphi(x)\geq 0$ for all $x\in R^\nu$, $\int \varphi(x)dx = 1$, $\varphi_t=x^\nu\varphi(\frac{x}{t})$, and f is a continuous function with compact support then $f*\varphi_t \to f$ uniformly as $t\to\infty$. Here $*$ indicates convolution. On the other hand $f*\varphi_t\in\mathcal{D}$, and $f*\varphi_t$ has a compact support independent of t for all $t\geq 1$. Hence $\mathcal{D}\subset S^c$ is dense in L_G^2 . Finally we recall the following result from the theory of distributions. The mapping $\varphi\to\tilde{\varphi}$ is an invertible bicontinuous transformation from S^c onto S^c . In particular, the functions $\tilde{\varphi},\varphi\in S^c$ are dense in L_G^2

Now we define the mapping

$$(3.1) \qquad I(\sum c_n \exp(i(n,x)) = \sum c_n X_n$$

in the discrete case, where the sum is finite, and

$$(3.1)' \qquad I(\widetilde{\varphi+i\psi}) = X(\varphi) + iX(\psi) , \quad \varphi,\psi \in S$$

in the generalized case.

Obviously

$$\|\sum c_n \exp[i(n,x)]\|_{L_G^2} = \sum\sum c_n \bar{c}_m \int \exp[i(n-m,x)]G(dx) =$$

$$= \sum\sum c_n \bar{c}_m \, E\, X_n X_m = E|\sum c_n X_n|^2 ,$$

and

$$\|\widetilde{\varphi+i\psi}\|_{L_G^2} = \int [|\tilde{\varphi}(x)|^2 + |\tilde{\psi}(x)|^2] G(dx) = E(|X(\varphi) + iX(\psi)|^2),$$

that is the mapping from a linear subspace of L_G^2 to H_1^c is norm preserving. Since the subspace where I is defined is dense in L_G^2, the mapping I can be uniquely extended to a norm preserving linear transformation from L_G^2 to H_1^c. Since the variables X_n or $X(\varphi)$, $\varphi \in S$, are obtained as the image of some element from L_G^2 under this transformation, I is a unitary transformation from L_G^2 onto H_1^c.

Now we define the random spectral measure $Z_G(A)$ for all $A \in B^{\nu}$ such that $G(A) < \infty$, by the formula

$$Z_G(A) = I(\chi_A)$$

where χ_A denotes the indicator function of the set A. It is clear that

(i) The random variables $Z_G(A)$ are complex valued jointly Gaussian random variables.

(ii) $E\, Z_G(A) = 0$

(iii) $E\, Z_G(A) \overline{Z_G(B)} = G(A \cap B)$

(iv) $\sum_{j=1}^{n} Z_G(A_j) = Z_G\left(\bigcup_{j=1}^{n} A_j\right)$ if $A_1, \ldots, A_n$ are disjoint

Also the relation

(v) $Z_G(A) = \overline{Z_G(-A)}$

holds true. This follows from the relation

(v)' $I(f) = \overline{I(f_-)}$ for all $f \in L_G^2$ where $f_-(x) = \overline{f(-x)}$.

Relation (v)' is obvious if f is a finite trigonometrical polynomial in the discrete case, or if $f = \tilde{\varphi}$, $\varphi \in S^c$ in the generalized case. Then a simple limiting procedure implies (v)' for all $f \in L_G^2$. Now we introduce the following

Definition

Let G be a spectral measure. A set of random variables $Z_G(A)$, $G(A) < \infty$, satisfying (i)-(v) is called a (Gaussian) random spectral measure correspoinding to G.

Given a Gaussian random spectral measure Z_G corresponding to a spectral measure G, we define the stochastic integral $\int f(x)\, Z_G(dx)$. Let us first consider simple functions of the form $f(x) = \sum c_i \chi_{A_i}(x)$, where the sum is finite and $G(A_i) < \infty$ for all i. Then we define

$$\int f(x)\, Z_G(dx) = \sum c_i\, Z_G(A_i)$$

Then

$$E\left|\int f(x)\, Z_G(dx)\right|^2 = \sum c_i \bar{c}_j\, G(A_i \cap A_j) = \int |f(x)|^2 G(dx). \tag{3.2}$$

Since the simple functions are dense in L_G^2, relation (3.2) enables us to define $\int f(x)\, Z_G(dx)$ for all $f \in L_G^2$ via L_2-continuity. For all stationary Gaussian fields with spectral measure G we have constructed a random spectral measure Z_G corresponding to G.

It is clear that

$$\int f(x)\, Z_G(dx) = I(f) \quad \text{for all} \quad f \in L_G^2$$

with this random spectral measure Z_G. In particular

(3.3) $$X_n = \int \exp[i(n,x)] \, Z_G(dx), \quad n \in \mathbb{Z}_\nu$$

in the discrete case, and

(3.3)' $$X(\varphi) = \int \tilde{\varphi}(x) \, Z_G(dx), \quad \varphi \in \mathcal{S}$$

in the generalized case. Now we formulate the following

Theorem 3.1

For a stationary Gaussian field (a discrete or a generalized one) with a spectral measure G, there exists a unique Gaussian random spectral measure Z_G corresponding to the spectral measure G on the same probability space as the Gaussian random field such that relation (3.3) or (3.3') holds in the discrete or generalized case, respectively.

Furthermore,

(3.4) $$\mathcal{B}(Z_G(A),\ G(A) < \infty) = \begin{cases} \mathcal{B}(X_n,\ n \in \mathbb{Z}_\nu) & \text{in the discrete case} \\ \mathcal{B}(X(\varphi),\ \varphi \in \mathcal{S}) & \text{in the generalized case} \end{cases}$$

We shall say that the random spectral measure Z_G satisfying Theorem 3.1 is adapted to the random field.

Proof of Theorem 3.1

We have already constructed a random spectral measure Z_G satisfying (3.3) or (3.3'). Since this random measure is measurable with respect to the random field, relations (3.3) or (3.3') imply (3.4) .

To prove the uniqueness, it is enough to observe that because of the linearity and L_2 continuity of stochastic integrals, relations (3.3) or (3.3') imply that

$$Z_G(A) = \int \chi_A(x)\, Z_G(dx) = I(\chi_A)$$

under the conditions of Theorem 3.1.

Finally we list some properties of Gaussian random spectral measures.

(vi) The random variables $\mathrm{Re}\, Z_G(A)$ are independent of the random variables $\mathrm{Im}\, Z_G(B)$.

(vii) If $A_1 \cup (-A_1), \dots, A_n \cup (-A_n)$ are disjoint, then the random variables $Z_G(A_1), \dots, Z_G(A_n)$ are independent.

(viii) If $A \cap (-A)$ is empty, then $Z_G(A)$ and $Z_G(-A)$ are independent with mean zero and variance $G(A)/2$.

These properties easily follow from (i)-(v). Since $Z_G(\cdot)$ are complex valued Gaussian random variables, to prove independence it is enough to show that the real and imaginary parts are independent. We show, as an example, the proof of (vi).

$$E\,\mathrm{Re}\, Z_G(A)\, \mathrm{Im}\, Z_G(B) = \frac{1}{4}[E(Z_G(A) + \overline{Z_G(A)})(Z_G(B) - \overline{Z_G(B)})] =$$

$$= \frac{1}{4} E[(Z_G(A) + Z_G(-A))(\overline{Z_G(-B)} - \overline{Z_G(B)})] =$$

$$= \frac{1}{4} [G(A \cap (-B)) - G(A \cap B) + G((-A) \cap (-B)) - G((-A) \cap B)] = 0 ,$$

since $G(D) = G(-D)$ for all $D \in \mathcal{B}^\nu$.
The properties of the random spectral measure Z_G listed above imply in particular, that the spectral measure G determines the joint disrtibution of the corresponding random variables $Z_G(B)$, $B \in \mathcal{B}^\nu$.

4. Multiple Wiener-Itô integrals

In this section we define the so-called multiple Wiener-Itô integrals, and we prove their most important properties with the help of Itô's formula, whose proof is postponed to the next section. More precisely we discuss in this section a modified version of the Wiener-Itô integrals, where we integrate with respect to a random spectral measure rather than a random measure with independent increments. This modification makes it necessary to slightly change the definition of the integral. This modified Wiener-Itô integral seems to be a more useful tool than the original one or the Wick polynomials because it enables us to describe the action of shift transformations.

Let G be the spectral measure of a stationary Gaussian field (discrete or generalized one.) We define the following real Hilbert spaces $\bar{H}_G^n$ and H_G^n , $n=1,2,\dots$. $f_n \in \bar{H}_G^n$ if and only if $f_n = f_n(x_1,\dots,x_n)$, $x_j \in R^\nu$, $j=1,2,\dots$ is a complex valued function of n variables, and

(a) $f_n(-x_1,\dots,-x_n) = \overline{f_n(x_1,\dots,x_n)}$

(b) $\| f_n \|^2 = \int |f_n(x_1,\dots,x_n)|^2 G(dx_1),\dots,G(dx_n) < \infty$.

(b) also define the norm in $\bar{H}^n_G$.

$H^n_G \subset \bar{H}^n_G$ contains those $f_n \in \bar{H}^n_G$ which are invariant under permutations of their arguments, i.e.

(c) $f_n(x_{\Pi(1)},\ldots,x_{\Pi(n)}) = f_n(x_1,\ldots,x_n)$ for all $\Pi \in \Pi_n$,

where Π_n denotes the group of all permutations of the set $\{1,2,\ldots,n\}$.

The norm in H^n_G is defined the same way as that in $\bar{H}^n_G$. We also define $H^o_G = \bar{H}^o_G$ to be the space of all real constants with the norm $\| c \| = |c|$.

We remark that $\bar{H}^n_G$ is actually the n-fold direct product of H^1_G, and H^n_G is the n-fold symmetrical direct product of H^1_G. Condition (a) means heuristically that f_n is the Fourier transform of a real valued function.

Finally we define the so called Fock space $ExpH_G$ whose elements are sequences $f=(f_o,f_1,\ldots)$, $f_n \in H^n_G$, such that

$$\| f \|^2 = \sum_{n=o}^{\infty} \frac{1}{n!} \| f_n \|^2 < \infty$$

Given an $f \in \bar{H}^n_G$, we define $\text{Sym}\, f$

$$\text{Sym}\, f(x_1,\ldots,x_n) = \frac{1}{n!} \sum_{\Pi \in \Pi_n} f(x_{\Pi(1)},\ldots,x_{\Pi(n)}) .$$

Clearly $\text{Sym}\, f \in H^n_G$, and

$$(4.1) \qquad \| \text{Sym}\, f \| \leq \| f \| .$$

Let Z_G be a Gaussian random spectral measure corresponding to G on a probability space $(\Omega,\mathcal{A},P)$. We shall define the n-fold Wiener-Itô integrals

$$\int f_n(x_1,\dots,x_n)\, Z_G(dx_1)\dots Z_G(dx_n)$$

for all $f_n \in \bar{H}_G^n$. Throughout this paper we shall use the notations

$$I_G(f_n) = \frac{1}{n!} \int f_n(x_1,\dots,x_n)\, Z_G(dx_1)\dots Z_G(dx_n) \quad \text{for} \quad f_n \in \bar{H}_G^n$$

and

$$I_G(f) = \sum_{n=0}^{\infty} I_G(f_n) \quad \text{for} \quad f=(f_0,\dots,f_1,\dots) \in \operatorname{Exp} H_G$$

We shall see that $I_G(f_n) = I_G\,(\operatorname{Sym} f_n)$ for all $f_n \in \bar{H}_G^n$. Therefore, it would have been sufficient to define the Wiener-Itô integral only for functions in H_G^n, and not for all functions in $\bar{H}_G^n$. Nevertheless, some arguments become simpler if we work in $\bar{H}_G^n$. In the definition of Wiener-Itô integrals first we restrict ourselves to the case when the spectral measure G is non-atomic, i.e. $G(\{x\}) = 0$ for all $x \in R^\nu$. This condition is satisfied in all interesting cases. However, we shall later show how one can get rid of this restriction.

First we define a subclass $\hat{\bar{H}}_G^n \subset \bar{H}_G^n$ of simple functions, and define the Wiener-Itô integral for those functions. Let $\mathcal{D} = \{\Delta_j,\ j=\pm 1, 2,\dots,\pm N\}$ be a finite set of rectangles in R^{ν}, indexed by the integers $\pm 1, \pm 2, \dots \pm N$. We say that $\mathcal{D}$ is a regular system of rectangles if $\Delta_j = -\Delta_{-j}$, and $\Delta_j \cap \Delta_\ell = \emptyset$ if $j \neq \ell$ for all $j, \ell = \pm 1, \pm 2, \dots, \pm N$. A function $f \in \bar{H}_G^n$ is adapted to this system $\mathcal{D}$ if $f(x_1, \dots, x_n)$ is constant on the sets $\Delta_{j_1} \times \Delta_{j_2} \times \dots \times \Delta_{j_n}$, $j_\ell = \pm 1, \dots, \pm N$, $\ell = 1, 2, \dots, n$, it vanishes outside these sets and also on the sets $\Delta_{j_1} \times \dots \times \Delta_{j_n}$ for which $j_\ell = \pm j_{\ell'}$ for some $\ell \neq \ell'$. A function $f \in \bar{H}_G^n$ is in the class $\hat{\bar{H}}_G^n$, of simple functions if it is adapted to some regular system of rectangles $\mathcal{D} = \{\Delta_j, j = \pm 1, \pm 2, \dots\}$, and its Wiener-Itô integral with respect to Z_G is defined as

$$\int f(x_1, \dots, x_n) Z_G(dx_1), \dots, Z_G(dx_n) = n!\, I_G(f) =$$

(4.2)

$$= \sum_{\substack{j_\ell = \pm 1, \dots, \pm N \\ \ell = 1, 2, \dots, n}} f(x_{j_1}, \dots, x_{j_n}) Z_G(\Delta_{j_1}) \dots Z_G(\Delta_{j_n}),$$

where $x_j \in \Delta_j$, $j = \pm 1, \dots, \pm N$. We remark that although the regular system $\mathcal{D}$ to which f is adapted, is not uniquely determined (the rectangles in $\mathcal{D}$ can be divided to smaller rectangles), the integral defined in (4.2) is meaningful, i.e. it does not depend on the choice of $\mathcal{D}$. All products $Z_G(\Delta_{j_1}), \dots, Z_G(\Delta_{j_n})$ with non-zero coefficient in (4.2) are products of independent random variables. We had this property in mind when requiring the condition that the function f vanishes on a rectangle $\Delta_{j_1} \times \dots \times \Delta_{j_n}$ if $j_\ell = \pm j_{\ell'}$, $\ell \neq \ell'$. This condition is interpreted in the literature as discarding the hyperplanes $x_\ell = x_{\ell'}$ and $x_\ell = -x_{\ell'}$, $\ell, \ell' = 1, 2, \dots, n$, from the domain of integration. Property (a) of the functions in $\bar{H}_G^n$ and property (v) of the random spectral measures imply that $I_G(f) = \overline{I_G(f)}$, i.e. $I_G(f)$ is a real valued random variable for all $f \in \bar{\hat{H}}_G^n$. The relation

$$E\, I_G(f) = 0, \qquad f \in \bar{\hat{H}}_G^n, \qquad n = 1, 2, \dots \tag{4.3}$$

also holds. Let $\hat{H}_G^n = H_G^n \cap \bar{\hat{H}}_G^n$. If $f \in \bar{\hat{H}}_G^n$ then $\mathrm{Sym}\, f \in \hat{H}_G^n$, and

$$I_G(f) = I_G(\mathrm{Sym}\, f) \tag{4.4}$$

Relation (4.4) follows immediately from the observation that $Z_G(\Delta_{j_1}) \dots Z_G(\Delta_{j_n}) = Z_G(\Delta_{\Pi(j_1)}) \dots Z_G(\Delta_{\Pi(j_n)})$ for all $\Pi \in \Pi_n$.

We also claim that

$$E\, I_G(f)^2 \leq \frac{1}{n!} \| f \|^2 \quad \text{for} \quad f \in \bar{\hat{H}}_G^n \tag{4.5}$$

and

$$E\, I_G(f)^2 = \frac{1}{n!} \| f \|^2 \quad \text{for} \quad f \in \hat{H}_G^n . \tag{4.5'}$$

Because of (4.1) and (4.4) it is enough to check (4.5')
Let $\mathcal{D}$ be a regular system of rectangles in R^{ν}, $j_1,\dots,j_n$ and $k_1,\dots,k_n$ be indices such that $j_\ell \neq \pm j_{\ell'}$, $k_\ell \neq \pm k_{\ell'}$ if $\ell \neq \ell'$. Then

$$E\, Z_G(\Delta_{j_1})\dots Z_G(\Delta_{j_n})\,\overline{Z_G(\Delta_{k_1})\dots Z_G(\Delta_{k_n})} =$$

$$= \begin{cases} G(\Delta_{j_1})\dots G(\Delta_{j_n}) & \text{if } \{j_1,\dots,j_n\} = \{k_1,\dots,k_n\} \\ 0 & \text{otherwise} \end{cases}$$

To see the last relation one has to observe that the product on the left-hand side can be written as a product of independent random variables, and if $\{j_1,\dots,j_n\} \neq \{k_1,\dots,k_n\}$ then at least one element of this product has zero expectation. (We remark that also $E Z_G(\Delta_j)\, Z_G(-\Delta_j)=0$). Therefore

$$E\, I_G(f)^2 = (\tfrac{1}{n!})^2 \sum\sum f(x_{j_1},\dots,x_{j_n})\,\overline{f(x_{k_1},\dots,x_{k_n})}$$

$$E\, Z_G(\Delta_{j_1})\dots Z_G(\Delta_{j_n}) Z_G(\Delta_{k_1})\dots Z_G(\Delta_{k_n}) =$$

$$= (\tfrac{1}{n!})^2 \sum |f(x_{j_1},\dots,x_{j_n})|^2\, G(\Delta_{j_1})\dots G(\Delta_{j_n})\, n! =$$

$$= \frac{1}{n!} \int |f(x_{j_1},\dots,x_{j_n})|^2\, G(dx_1)\dots G(dx_n) = \frac{1}{n!}\, \|f\|^2 .$$

If $f\in\hat{\bar{H}}_G^n$, $f'\in\hat{\bar{H}}_G^{n'}$ then a regular system $\mathcal{D}$ of rectangles be found in such a way that both f and f' are adapted to $\mathcal{D}$. Hence an argument similar to the proof of (4.5') shows that

$$(4.6) \qquad E\, I_G(f) I_G(f')=0 \quad \text{if} \quad f\in\hat{\bar{H}}_G^n, \quad f'\in\hat{\bar{H}}_G^{n'}, \quad n\neq n' .$$

We show that $\hat{\bar{H}}_G^n$ is dense in $\bar{H}_G^n$ (and $\hat{H}_G^n$ is dense in H_G^n). Let $A=D_1 x\ldots x D_n$ be the direct product of rectangles from R^ν, and let $A\cap(-A)=\emptyset$. It is enough to show that the indicator function of $A\cap(-A)$ can be approximated arbitrary well by functions from $\hat{\bar{H}}_G^n$. Let $M=\max G(D_j)$. A regular system of rectangles in R^ν $\mathcal{D}=\{\Delta_j,\ j=\pm 1,\ldots,\pm M\}$ can be chosen in such a way that all sets D_j and $-D_j$ can be expressed as the union of some Δ from $\mathcal{D}$. Moreover we may assume that $G(\Delta)<\varepsilon$ for all $\Delta\in\mathcal{D}$. Indeed, $A\cap(-A)$ can be covered with finitely many rectangles with G measure less than ε by the non-atomic property of G and the Heyne-Borel theorem. Then a regular system $\mathcal{D}$ can be found with the required properties, and such that all $\Delta\in\mathcal{D}$ is contained in one of these covering rectangles. (This is the point where we have exploited that G is non-atomic). Let us write $A\cup(-A)$ as the union of rectangles of the form $\Delta_{j_1} x\ldots x\Delta_{j_n}$, and let us discard the products for which $j_\ell=\pm j_{\ell'}$, with some $\ell\neq\ell'$. The indicator function of the union of the remaining products is in $\hat{\bar{H}}_G^n$, and it approximates the function $\chi_{A\cup(-A)}(x)$ very well, since for all pairs $\ell,\ell',\ell\neq\ell'$, the sum of the terms $G(\Delta_{j_1})\dots G(\Delta_{j_n})$ for which $j_\ell=\pm j_{\ell'}$ is less than $2\varepsilon M^{n-1}$.

As the transformation $I_G(f)$ is a contraction from $\hat{H}_G^n$ into $L_2(\Omega,A,P)$, it can uniquely be extended to the closure of $\hat{H}_G^n$, i.e. to $\bar{H}_G^n$. We define the n-fold Wiener-Itô integral in the general case via this extension. $I_G(f)$ is a real valued random variable for all $f\in\bar{H}_G^n$, and relations (4.3), (4.5), (4.5') remain valid for $f,f'\in\bar{H}_G^n$ or $f\in H_G^n$ instead of $f,f'\in\hat{H}_G^n$ or $f\in\hat{H}_G^n$. For $f=(f_0,f_1,\dots)\in Exp\ H_G$ we define $I_G(f)=\sum I_G(f_n)$. Relations (4.5') and (4.6) imply that the transformation $I_G\colon Exp\ H_G\to L^2(\Omega,A,P)$ is an isometry. We shall show that also the following result holds true:

Theorem 4.1.

Let a stationary Gaussian random field be given (discrete or generalized one), and let Z_G denote the random spectral measure adapted to it. If we integrate with respect to this Z_G then the transformation $I_G\colon \exp H_G\to H$ is unitary. The transformation $(n!)^{1/2}I_G\colon H_G^n\to H_n$ is also unitary.

In the proof we need an identity whose proof is postponed to the next section.

Theorem 4.2. Itô's formula

Let $\varphi_1, \varphi_2, \ldots, \varphi_m; \varphi_j \in H_G^1$, *be an orthonormal system in* L_G^2. *Let some positive integers* $j_1, \ldots, j_m$ *be given,* $j_1 + \ldots + j_m = N$, *and define, for all* $i = 1, 2, \ldots, N$, *the functions* $g_i = \varphi_s$ *for* $j_1 + \ldots + j_{s-1} < i \le j_1 + \ldots + j_s$. *Then*

$$H_{j_1}\left(\int \varphi_1(x) Z_G(dx)\right) \ldots H_{j_m}\left(\int \varphi_m(x) Z_G(dx)\right) =$$

$$= \int g_1(x_1) \ldots g_N(x_N) Z_G(dx_1) \ldots Z_G(dx_N) =$$

$$= \int \mathrm{Sym}[g_1(x_1) \ldots g_N(x_N)] Z_G(dx_1) \ldots Z_G(dx_N) .$$

($H_j(x)$ denotes again the j-th Hermite polynomial with leading coefficient 1.)

Proof of Theorem 4.1.

The one-fold integral $I_G(f)$, $f \in H_G^1$, agrees with the stochastic integral $I(f)$ defined in section 3. Hence $I_G(\exp(i(n,x))) = X_n$ in the discrete case, and $I_G(\tilde{\varphi}) = X(\varphi)$, $\varphi \in S$. Hence $I_G : H_G^1 \to \mathcal{H}_1$ is a unitary transformation. Let $\varphi_1, \varphi_2, \ldots$ be a complete orthonormal basis in H_G^1. Then $\xi_j = \int \varphi_j(x)\, Z_G(dx)$, $j = 1, 2, \ldots$ is a complete orthonormal basis in $\mathcal{H}_1$. Itô's formula implies that $H_{j_1}(\xi_1) \ldots H_{j_m}(\xi_m)$ can be written as a $(j_1 + \ldots + j_m)$-fold integral. Therefore Theorem 2.1 implies that the image of $\exp H_G$ is the whole space $\mathcal{H}$, and $I_G : \mathrm{Exp}\, H_G \to \mathcal{H}$ is unitary.

The image of H_G^n contains H_n because of Corollary 2.4 and Itô's formula. Since these images are orthogonal for different n, formula (2.1) implies that the image of H_G^n coincides with H_n. Hence $(n!)^{1/2} I_G : H_G^n \to H_n$ is a unitary transformation.

The next result describes the action of shift transformations in H. We know by Theorem 4.1 that all $\eta \in H$ can be written in the form

$$(4.7) \quad \eta = f_o + \sum_{n=1}^{\infty} \frac{1}{n!} \int f_n(x_1,\dots,x_n)\, Z_G(dx_1)\dots Z_G(dx_n)$$

with $f=(f_o, f_1, \dots) \in Exp\, H_G$ in a unique way, where Z_G is the random measure adapted to the stationary Gaussian field.

Theorem 4.3

Let $\eta \in H$ have the form (4.7). Then

$$T_t \eta = f_o + \sum_{n=1}^{\infty} \frac{1}{n!} \int \exp[i(t, x_1+\dots+x_n)] f_n(x_1,\dots,x_n)$$

$$Z_G(dx_1)\dots Z_G(dx_n)$$

for all $t \in R^\nu$ in the generalized and for all $t \in Z_\nu$ in the discrete case.

Proof of Theorem 4.3

The relations

$$T_t \int e^{i(n,x)} Z_G(dx) = X_{n+t} = \int e^{i(t,x)}\, e^{i(n,x)} Z_G(dx)$$

and

$$T_t \int \varphi(x) Z_G(dx) = X(T_t\varphi) = \int e^{i(t,x)} \varphi(x) Z_G(dx), \quad \varphi \in S$$

holds in the discrete and generalized case, respectively. These relations imply that Theorem 4.3 holds true in the special case when η is a one-fold integral. Let $f_1(x),\dots,f_m(x)$ be an orthonormal system in H_G^1. Then $e^{i(t,x)} f_1(x),\dots,e^{i(t,x)} f_m(x)$ is also an orthonormal system in H_G^1. Hence Itô's formula implies that Theorem 4.3 holds also for $\eta = H_{j_1}(\int f_1(x) Z_G(dx))\dots$ $\dots H_{j_m}(\int f_m(x) Z_G(dx))$. As the linear combinations of such η are dense in H, Theorem 4.3 holds true. The next result is a formula for the change of variables in Wiener-Itô integrals.

Theorem 4.4

Let G and G' be non-atomic spectral measures such that G is absolutely continuous with respect to the measure G' and let $g(x)$ be a complex valued function such that

$$g(x) = \overline{g(-x)}$$
$$|g(x)|^2 = \frac{dG(x)}{dG'(x)} .$$

For every $f=(f_o, f_1,\dots) \in \operatorname{Exp} H_G$ we define

$$f'_n(x_1,\dots,x_n) = f_n(x_1,\dots,x_n) g(x_1)\dots g(x_n), \quad n=1,2,\dots, \quad f'_o = f_o .$$

Then $f' = (f'_o, f'_1, \dots) \in \operatorname{Exp} H_{G'}$, and

$$f_o + \sum \frac{1}{n!} \int f_n(x_1,\dots,x_n) Z_G(dx_1)\dots Z_G(dx_n) \triangleq f'_o +$$

$$+ \sum \frac{1}{n!} \int f'_n(x_1,\dots,x_n)\, Z_{G'}(dx_1)\dots Z_{G'}(dx_n)$$

<u>where</u> Z_G <u>and</u> $Z_{G'}$ <u>are Gaussian random spectral measures corresponding to</u> G <u>and</u> G'.

<u>Proof of Theorem 4.4</u>

$\| f'_n\|_{L^2_{G'}} = \| f_n\|_{L^2_G}$, hence $f' \in Exp\, H_{G'}$. Let $\varphi_1, \varphi_2, \dots$ be a complete orthonormal system in H^1_G. Then $\varphi'_1, \varphi'_2, \dots; \varphi'_j(x) = \varphi_j(x) g(x)$, is a complete orthonormal system in H^1_G. All functions $f_n \in H^n_G$ can be written in the form $f(x_1,\dots,x_n) = \sum c_{j_1,\dots,j_n}$ Sym$(\varphi_{j_1}(x_1)\dots$ $\dots\varphi_{j_n}(x_n))$. Then $f'_n(x_1,\dots,x_n) = \sum c_{j_1,\dots,j_n}$ Sym $(\varphi'_{j_1}(x_1)\dots\varphi'_{j_n}(x_n))$. Rewriting all terms $\int \text{Sym}[\varphi_{j_1}(x_1)\dots\varphi_{j_n}(x_n)]\, Z_G(dx_1)\dots Z_G(dx_n)$ and $\int \text{Sym}[\varphi'_{j_1}(x_1)\dots\varphi'_{j_n}(x_n)]\, Z_{G'}(dx_1)\dots Z_{G'}(dx_n)$ by means of Itô's formula we get that f and f' depend on a sequence of independent standard normal variables in the same way. Theorem 4.4 is proven.

The next result makes a connection between Wick polynomials and Wiener-Itô integrals.

<u>Theorem 4.5</u>

<u>Let a stationary Gaussian field be given, and let</u> Z_G <u>denote the random spectral measure adapted to it. Let</u>

$P(x_1,\dots,x_m)$ <u>be a homogeneous polynomial of degree</u> n, <u>and let</u> $h_1,\dots,h_m \in H_G^1$. <u>Define</u> $\xi_j = \int h_j(x) Z_G(dx)$, $j=1,2,\dots,m$ <u>and</u> $\bar{P}(x_1,\dots,x_n) = \sum c_{j_1,\dots,j_n} h_{j_1}(x_1)\dots$ $\dots h_{j_n}(x_n)$, <u>where</u> $P(x_1,\dots,x_m) = \sum c_{j_1,\dots,j_n} x_{j_1}\dots x_{j_n}$ <u>Then</u> $:P(\xi_1,\dots,\xi_m): = \int \bar{P}(x_1,\dots,x_n)\, Z_G(dx_1)\dots Z_G(dx_n)$.

<u>Remark</u>

If P is a polynomial of degree n then it can be written as $P=P_1+P_2$, where P_1 is a homogeneous polynomial of degree n, and P_2 is a polynomial of degree less than n. Obviously

$$:P(\xi_1,\dots,\xi_m): = :P_1(\xi_1,\dots,\xi_m): \quad .$$

<u>Proof of Theorem 4.5</u>

It is enough to show that

$$:\xi_{j_1}\dots\xi_{j_n}: = \int h_{j_1}(x_1)\dots h_{j_n}(x_n) Z_G(dx_1)\dots Z_G(dx_n)$$

If $h_1,\dots,h_m \in H_G^1$ are orthogonal this relation follows from a comparison of Corollary 2.3 with Itô's formula. In the general case an orthogonal system $\bar{h}_1,\dots,\bar{h}_m \in H_G^1$ can be found such that

$$h_j = \sum c_{j,k}\, \bar{h}_k,\ j=1,\dots,m$$

with some real constants $c_{j,k}$. Set $\eta_k = \int \bar{h}_j(x) Z_G(dx)$.

Then

$$:\xi_{j_1}\dots\xi_{j_n}: = :(\sum c_{j_1,k}\eta_k)\dots(\sum c_{j_n,k}\eta_k): =$$

$$= \sum c_{j_1,k_1}\dots c_{j_n,k_n} :\eta_{k_1}\dots\eta_{k_n}: =$$

$$= \sum c_{j_1,k_1}\dots c_{j_n,k_n} \int \bar{h}_{k_1}(x_1)\dots\bar{h}_{k_n}(x_1)\dots\bar{h}_{k_n}(x_n)Z_G(dx_1)\dots$$

$$\dots Z_G(dx_n) = \int h_{j_1}(x_1)\dots h_{j_n}(x_n)Z_G(dx_1)\dots Z_G(dx_n)$$

as claimed.

We finish this section by showing how the Wiener-Itô integral can be defined if the spectral measure G may have atoms. If we try to give this definition by modifying the original one, then we have to split up the atoms. The simplest way we found for this splitting up, was the use of randomization.

Let G be a spectral measure on R^ν, and let Z_G be a corresponding Gaussian spectral random measure on a probability space $(\Omega,\mathcal{A},P)$. Let us define a new spectral measure $\hat{G} = G\times\lambda_{[-\frac{1}{2},\frac{1}{2}]}$ on $R^{\nu+1}$, where $\lambda_{[-\frac{1}{2},\frac{1}{2}]}$ denotes the uniform distribution on the interval $[-\frac{1}{2},\frac{1}{2}]$. If the probability space $(\Omega,\mathcal{A},P)$ is sufficiently rich, a random spectral measure $Z_{\hat{G}}$ corresponding to $\hat{G}$ can be defined on it in such a way that $Z_{\hat{G}}(A\times[-\frac{1}{2},\frac{1}{2}])=Z_G(A)$ for all $A\in\mathcal{B}^\nu$. For $f\in\bar{H}_G^n$ we define the function $\hat{f}\in\bar{H}_{\hat{G}}^n$ by the formula $\hat{f}(y_1,\dots,y_n)=f(x_1,\dots,x_n)$ if y_j is the juxtaposition (x_j,u_j), $x_j\in R^\nu$, $u_j\in R^1$, $j=1,2,\dots,n$. Finally we define the Wiener-Itô integral in the general case by the formula

$$\int f(x_1,\dots,x_n) Z_G(dx_1)\dots Z_G(dx_n) =$$

$$= \int \hat{f}(y_1,\dots,y_n)\ Z_{\hat{G}}(dy_1)\dots Z_{\hat{G}}(dy_n)$$

(What we actually have done was to introduce a virtual new coordinate u. With the help of this new coordinate we could reduce the general case to the special case when G is non-atomic.) If G is a non-atomic spectral measure then the new definition of Wiener-Itô integrals coincides with the original one. It is easy to check this fact for one-fold integrals, and then Itô's formula proves it for multiple integrals. It can be seen, with the help of Itô's formula again, that all results in this section remain valid for the new definition of Wiener-Itô integrals. In particular, we formulate the following result.

Given a stationary Gaussian field let Z_G be the random spectral measure adapted to it. All $f \in H_G^n$ can be written in the form

$$f(x_1,\dots,x_n) = \sum c_{j_1,\dots,j_n} \varphi_{j_1}(x_1)\dots\varphi_{j_n}(x_n) \tag{4.8}$$

with some $\varphi_j \in H_G^1$, $j=1,2,\dots$. Define $\xi_j = \int \varphi_j(x)\ Z_G(dx)$. If f has the form (4.8) then

$$\int f(x_1,\dots,x_n) Z_G(dx_1)\dots Z_G(dx_n) = \sum c_{j_1,\dots,j_n} :\xi_{j_1}\dots\xi_{j_n}: .$$

The last identity would provide another possibility for defining Wiener-Itô integrals.

5. The proof of Itô's formula. The diagram formula and some of its consequences

We shall prove Itô's formula with the help of the following

Proposition 5.1

Let $f \in \bar{H}_G^n$ and $h \in \bar{H}_G^1$. Let us define the functions

$$f \underset{k}{\times} h(x_1,\dots,x_{k-1}, x_{k+1},\dots,x_n) =$$

$$= \int f(x_1,\dots,x_n)\, \overline{h(x_k)}\, G(dx_k)\,, \qquad k=1,2,\dots,n$$

and

$$fh(x_1,\dots,x_{n+1}) = f(x_1,\dots,x_n)h(x_{n+1})\,.$$

Then $f \underset{k}{\times} h$, $k=1,2,\dots,n$, *and* fh *are in* $\bar{H}_G^{n-1}$ *and* $\bar{H}_G^{n+1}$, *respectively and their norms satisfy the inequality* $\| f \underset{k}{\times} h \| \le \| f \| \cdot \| h \|$, $\| fh \| \le \| f \| \cdot \| h \|$

The relation

$$nI_G(f)I_G(h) = n(n+1)I_G(fh) + \sum_{k=1}^{n} I_G(f \underset{k}{\times} h)$$

holds true.

We also need the following recursion formula for Hermite polynomials.

Lemma 5.2

$H_n(x) = xH_{n-1}(x)-(n-1)H_{n-2}(x)$ for $n=1,2,\ldots$.

Proof of Lemma 5.2

$$H_n(x) = (-1)^n \exp(\frac{x^2}{2}) \cdot \frac{d^n}{dx^n} (\exp(-\frac{x^2}{2})) =$$

$$= -\exp(\frac{x^2}{2})\frac{d}{dx} [H_{n-1}(x)\exp(-\frac{x^2}{2})] = x\, H_{n-1}(x)-\frac{d}{dx} H_{n-1}(x)$$

Since $\frac{d}{dx} H_{n-1}(x)$ is a polynomial of order $n-2$ with leading coefficient $n-1$, we can write

$$\frac{d}{dx} H_{n-1}(x) = (n-1)H_{n-2}(x) + \sum_{j=o}^{n-3} c_j\, H_j(x) .$$

To complete the proof of Lemma 5.2 it remains to show that all coefficients c_j are zero. This follows from the orthogonality of the Hermite polynomials and the calculation

$$\int e^{-\frac{x^2}{2}} H_j(x) \frac{d}{dx} H_{n-1}(x)dx = -\int H_{n-1}(x)\frac{d}{dx}[H_j(x)e^{-\frac{x^2}{2}}]\, dx =$$

$$= \int e^{-\frac{x^2}{2}} H_{n-1}(x)H_{j+1}(x)dx = 0$$

for $j \leq n-3$.

Proof of Theorem 4.2 via Proposition 5.1

We prove Theorem 4.2 by induction.

Theorem 4.2 holds for $N=1$. Assume it holds for $N-1$. Let us define the functions

$$f(x_1,\ldots,x_{N-1}) = g_1(x_1)\ldots g_{N-1}(x_{N-1})$$
$$h(x) = g_N(x)$$

Then

$$J = \int g_1(x_1)\ldots g_N(x_N)Z_G(dx_1)\ldots Z_G(dx_N) = N!\; I_G(fh) =$$

$$= (N-1)!\; I_G(f)I_G(h) - \sum_{k=1}^{N-1} (N-2)!\; I_G(f \overset{x}{k} h)$$

by the Proposition 5.1. The induction hypothesis implies that

$$J = H_{j_1}(\int \varphi_1(x)Z_G(dx))\ldots$$

$$\ldots H_{j_{m-1}}(\int \varphi_{m-1}(x)Z_G(dx))H_{j_m-1}(\int \varphi_{m-1}(x)Z_G(dx))$$

$$\int \varphi_m(x)Z_G(dx) - (j_m-1)H_{j_1}(\int \varphi_1(x)Z_G(dx))\ldots$$

$$\ldots H_{j_{m-1}}(\int \varphi_{m-1}(x)Z_G(dx)) \quad H_{j_m-2}(\int \varphi_m(x)Z_G(dx)) ,$$

since

$$f \underset{k}{\times} h = \int g_1(x_1)\dots g_{N-1}(x_{N-1})\overline{\varphi_m(x_k)}G(dx_k) =$$

$$= \begin{cases} 0 & \text{if} \quad k \le N-j_m \\ g_1(x_1)\dots g_{k-1}(x_{k-1})g_{k+1}(x_{k+1})\dots g_{N-1}(x_{N-1}) & \text{if} \quad N-j_m < k \le N-1 \quad . \end{cases}$$

Hence Lemma 5.2 implies that

$$J = \prod_{s=1}^{m-1} H_{j_s}\left(\int \varphi_s(x)Z_G(dx)\right)\left[H_{j_m-1}\left(\int \varphi_m(x)Z_G(dx)\right)\int \varphi_m(x)Z_G(dx) - \right.$$

$$\left. - (j_m-1)H_{j_m-2}\left(\int \varphi_m(x)Z_G(dx)\right)\right] = \prod_{s=1}^{m} H_{j_s}\left(\int \varphi_s(x)Z_G(dx)\right) ,$$

as claimed.

Let us fix some functions $h_1 \in \bar{H}_G^{n_1}, \dots, h_m \in \bar{H}_G^{n_m}$. In the next result, in the so-called diagram formula, we express the product $I_G(h_1)\dots I_G(h_m)$ as the sum of Wiener-Itô integrals. This result contains Proposition 5.1 as a special case. There is no unique terminology for this result in the literature. We shall follow the notation of Dobrushin in [6].

We shall use the term diagram of order $(n_1,\dots,n_m)$ for an undirected graph of $n_1+\dots+n_m$ vertices such that its vertices are indexed by the pairs of integers (j,ℓ) , $\ell=1,\dots,m$, $j=1,\dots,n_\ell$, such that no more than one edge enters into each vertex, and such that edges can connect only pairs of vertices (j_1,ℓ_1), (j_2,ℓ_2) for which $\ell_1 \neq \ell_2$. Let $\Gamma=\Gamma(n_1,\ldots,n_m)$ denote the set of all diagrams. Given a $\gamma\in\Gamma$, $|\gamma|$ denotes the number of edges in γ. Let there be given a set of functions $h_1\in\bar{H}_G^{n_1},\ldots,h_m\in\bar{H}_G^{n_m}$. Write $N=n_1+\ldots+n_m$. We introduce the function of N variables $x_{j,\ell}$ corresponding to the vertices of the diagram by the formula

$$h(x_{j,\ell}\,,\ \ell=1,\ldots,m,\ \ j=1,\ldots,n_\ell) =$$

$$= \prod_{\ell=1}^{n} h(x_{j,\ell}\,,\ \ j=1,\ldots,n_\ell)$$

Fixing a diagram $\gamma\in\Gamma$ we enumerate the variables $x_{j,\ell}$ in such a way that the vertices into which no edges enter will have the numbers $1,2,\ldots,N-2|\gamma|$ and the vertices connected by an edge will have the numbers p and $p+|\gamma|$, where $p=N-2|\gamma|+1,\ldots,N-|\gamma|$. Let

$$h_\gamma(x_1,\ldots,x_{N-2|\gamma|}) =$$

$$= \int\ldots\int h(x_1,\ldots,x_{N-\gamma},\ -x_{N-2\gamma+1},\ldots,-x_{N-\gamma})$$

$$G(dx_{N-2|\gamma|+1})\ldots G(dx_{N-|\gamma|})\ .$$

The function h_γ depends only on the variables $x_1,\ldots,x_{N-2|\gamma|}$, i.e. it is independent of how the vertices connected by edges are indexed. Indeed, it follows from the evenness of the spectral measure that by interchanging the indices s and $s+|\gamma|$ of two vertices connected by an edge one does not change the value of the integral h_γ. Let us now consider $I_G(h_\gamma)$. The function h_γ may depend on the numbering of those vertices of γ from which no edge starts, but $\mathrm{Sym}\, h_\gamma$ and therefore $I_G(h_\gamma)$ does not depend on it. Now we formulate the following

Theorem 5.3

For any $h_1 \in \bar{H}_G^{n_1}, \dots, h_m \in \bar{H}_G^{n_m}$, $n_1, \dots, n_m = 1, 2, \dots$ the following relations hold true:

A) $h_\gamma \in \bar{H}_G^{N-2|\gamma|}$, and $\| h_\gamma \| \le \prod_{j=1}^{m} \| h_j \|$ for all $\gamma \in \Gamma$

B) $I_G(h_1) \dots I_G(h_m) = \sum_{\gamma \in \Gamma} \frac{(N-2|\gamma|)!}{n_1! \dots n_m!} I_G(h_\gamma)$.

Remark

In the special case $m=2$, $n_1=n$, $n_2=1$ Theorem 5.3 concides with Proposition 5.1. To see this it is enough to observe that $h(-x) = \overline{h(x)}$ for all $h \in \bar{H}_G^1$.

Proof of Theorem 5.3

It suffices to prove Theorem 5.2 in the special case $m=2$. Then the case $m>2$ follows by induction. We shall use the notation $n_1=n$, $n_2=m$, and we write $x_1, \dots, x_{n+m}$ instead of $x_{(1,1)}, \dots, x_{(1,n)}, x_{(2,1)}, \dots, x_{(2,m)}$.

Part A of Theorem 5.3 is a consequence of the Schwartz inequality. It is clear that h_γ satisfies property a) of the classes $\bar{H}_G^j$ defined in Section 4. To prove the estimate on the norm of h_γ it is enough to restrict ourselves to diagrams for which the vertices $(1,n)$ and $(2,m)$; $(1,n-1)$ and $(2,m-1)$; ...; $(1,n-k)$ and $(2,m-k)$ are connected by edges $0\le k\le \min(n,m)$. We can write

$$\Big|\int h_1(x_1,\dots,x_n)h_2(x_{n+1},\dots,x_{n+m-k-1},-x_{n-k},\dots$$

$$\dots,-x_n)G(dx_{n-k})\dots G(dx_n)\Big|^2 \le \int |h_1(x_1,\dots,x_n)|^2 G(dx_{n-k})\dots$$

$$\dots G(dx_n) \int |h_2(x_{n+1},\dots,x_{n+m})|^2\, G(dx_{n+m-k})\dots G(dx_{n+m})$$

by the Schwartz inequality. Integrating this inequality with respect to its free variables we get part A) of Theorem 5.3.

In the proof of part B) first we restrict ourselves to the case when $h_1\in\hat{H}_G^n$, $h_2\in\hat{H}_G^m$. Assume they are adapted to a regular system of $\mathcal{D}=\{\Delta_j,\ j=\pm 1,\dots,\pm N\}$ of rectangles R^ν. We may assume that all $\Delta_j\in\mathcal{D}$ satisfy $G(\Delta_j)<\varepsilon$ with a sufficiently small $\varepsilon>0$ to be chosen later, because otherwise we could split up the rectangles Δ_j into smaller ones. Let $K_i=\sup_x|h_i(x)|$ $i=1,2$, and let A be a cube containing all Δ_j.

We can write

$$I=I_G(h_1)I_G(h_2) = \frac{1}{n!m!}\sum{}' h_1(x_{j_1},\dots,x_{j_n})h_2(x_{k_1},\dots,x_{k_m})$$

$$Z_G(\Delta_{j_1})\dots Z_G(\Delta_{j_n})Z_G(\Delta_{k_1})\dots Z_G(\Delta_{k_m}),$$

where the summation in $\sum{}'$ goes through all pairs j_p, $k_r\in\{\pm 1,\dots,\pm N\}$, $p=1,\dots,n$, $r=1,\dots,m$, such that $j_p\neq\pm j_{\bar p}$, $k_r\neq\pm k_{\bar r}$, if $p\neq\pm\bar p$ or $r\neq\pm\bar r$, and $x_{j_p}\in\Delta_{j_p}$, $x_{k_r}\in\Delta_{k_r}$. Write

$$I = \sum_{\gamma\in\Gamma}\frac{1}{n!m!}\sum{}^{\gamma} h_1(x_{j_1},\dots,x_{j_n})h_2(x_{k_1},\dots,x_{k_m})Z_G(\Delta_{j_1})\dots$$

$$\dots Z_G(\Delta_{j_n})\ Z_G(\Delta_{k_1})\dots Z_G(\Delta_{k_m}) ,$$

where $\sum^{\gamma}$ contains those terms of $\sum{}'$ for which $j_p=k_r$ or $j_p=-k_r$ if the vertices $(1,p)$ and $(2,r)$ are connected in γ, and $j_p\neq\pm k_r$ otherwise. Let us introduce the notations $\underline{n}=\{1,2,\dots,n\}$, $\underline{m}=\{1,2,\dots,m\}$, $A_1=A_1(\gamma) = \{p\,|\,p\in\underline{n}$, and no edge starts from $(1,p)$ in $\gamma\}$, $A_2=A_2(\gamma) = \{r\,|\,r\in\underline{m}$, and no edge starts from $(2,r)$ in $\gamma\}$, and $B=B(\gamma) = \{(p,r),\ p\in\underline{n},\ r\in\underline{m}$, $(1,p)$ and $(2,r)$ are connected in $\gamma\}$.

Write

$$\sum{}^{\gamma} = \sum{}_1^{\gamma} + \sum{}_2^{\gamma}$$

with

$$\sum{}_1^{\gamma} = \sum{}^{\gamma} h_1(x_{j_1},\dots,x_{j_n})h_2(x_{k_1},\dots,x_{k_r})$$

$$\prod_{(p,r)\in B} E((Z_G(\Delta_{j_p})Z_G(\Delta_{k_r})) \prod_{p\in A_1} Z_G(\Delta_{j_p}) \prod_{r\in A_2} Z_G(\Delta_{k_r})$$

and

$$\sum_2^{\gamma} = \sum^{\gamma} h_1(x_{j_1},\dots,x_{j_n})h_2(x_{k_1},\dots,x_{k_r}) \prod_{p\in A_1} Z_G(\Delta_{j_p}) \prod_{r\in A_2} Z_G(\Delta_{k_r})\cdot$$

$$\cdot\Big[\prod_{(p,r)\in B} Z_G(\Delta_{j_p})Z_G(\Delta_{k_r}) - E\Big(\prod_{(p,r)\in B} Z_G(\Delta_{j_p})Z_G(\Delta_{k_r})\Big)\Big] .$$

We are going to show that $\sum_1^{\gamma}$ is a good approximation of $(n+m-2\gamma)!I_G(h_\gamma)$, and $\sum_2^{\gamma}$ is negligible. Since

$$EZ_G(\Delta_{j_p})Z_G(\Delta_{k_r}) = \begin{cases} G(\Delta_{j_p}) & \text{if } j_p=-k_r \\ 0 & \text{if } j_p=k_r \end{cases}$$

$$\sum_1^{\gamma} = (m+n-2|\gamma|)!\; I_G(h_{\gamma,\mathcal{D}}) ,$$

where

$$h_{\gamma,\mathcal{D}}(x_1,\dots,x_{n+m-2|\gamma|}) = \begin{cases} h_\gamma(x_1,\dots,x_{n+m-2|\gamma|}) \text{ if } x_j\in\Delta_{k_j}, \\ j=1,2,\dots,n+m-2|\gamma|, \text{ and the sets} \\ \Delta_{k_1},-\Delta_{k_1},\dots,\Delta_{k_{m+n-2|\gamma|}},-\Delta_{k_{m+n-2|\gamma|}} \\ \text{are different} \\ 0 \quad \text{otherwise} \end{cases}$$

(The difference between h_γ and $h_{\gamma,\mathcal{D}}$ is that h_γ vanishes not necessarily for $\Delta_{k_j}=\pm\Delta_{k_{\bar{j}}}$ with $1\le j\le n-|\gamma|$, $n-|\gamma|+1\le\bar{j}\le n+m-2|\gamma|$. The function $h_{\gamma,\mathcal{D}}\in\hat{\bar{H}}_G^{n+m-2|\gamma|}$, but h_γ needs not belong to this class.)

$$\sup|h_{\gamma,\mathcal{D}}| \le K_1 K_2 L^{|\gamma|} ,$$

and

$$E[(m+n-2|\gamma|)!\ I_G(h_\gamma) - \Sigma_1^\gamma]^2 \le$$

(5.1)

$$\le (m+n-2|\gamma|)!\ \| h_\gamma - h_{\gamma,\mathcal{D}} \|^2 \le (m+n-2\gamma)!\ 2mnK_1^2\ K_2^2\ L^{\max(m,n)}\varepsilon^d$$

if $d=\min(n,m)-|\gamma| > 0$, and

(5.1)' $$h_{\gamma,\mathcal{D}} = h_\gamma$$

if $d=0$.

Now we turn to the estimation of $E(\Sigma_2^\gamma)^2$. Let us express $(\Sigma_2^\gamma)^2$ as a sum by carrying out the multiplications in $(\Sigma_2^\gamma)^2$. We want to show that most terms of $(\Sigma_2^\gamma)^2$ have an expectation zero and the contribution of the remaining terms is negligible. We have to investigate expressions of the following form:

$$E\{ \prod_{p\in A_1} Z_G(\Delta_{j_p}) \prod_{r\in A_2} Z_G(\Delta_{k_r}) \prod_{\bar p\in A_1} Z_G(\Delta_{j_{\bar p}}) \prod_{\bar r\in A_2} Z_G(\Delta_{k_{\bar r}})$$

(5.2) $$[\prod_{(p,r)\in B} Z_G(\Delta_{j_p}) Z_G(\Delta_{k_r}) - E \prod_{(p,r)\in B} Z_G(\Delta_{j_p}) Z_G(\Delta_{k_r})]$$

$$[\prod_{(\bar p,\bar r)\in B} Z_G(\Delta_{j_{\bar p}}) Z_G(\Delta_{k_{\bar r}}) - E \prod_{(\bar p,\bar r)\in B} Z_G(\Delta_{j_{\bar p}}) Z_G(\Delta_{k_{\bar r}})]\} ,$$

where the indices $j_p, k_r, j_{\bar p}, k_{\bar r}$ are in the set $\{\pm 1,\dots,\pm N\}$, $j_p=k_r$ or $j_p=-k_r$ if $(p,r)\in B$, the numbers

$\pm j_p$, $\pm k_r$ are all different otherwise, and the same relations hold for $j_{\bar{p}}$ and $k_{\bar{r}}$. Define the sets

$$A=\{j_p,\ p\in A_1\} \cup \{k_r,\ r\in A_2\} \quad \text{and} \quad \bar{A}=\{j_{\bar{p}},\ \bar{p}\in A_1\} \cup \{k_{\bar{r}},\ \bar{r}\in A_2\}.$$

We claim that the expression (5.2) equals zero if $\bar{A}\neq -A$. Indeed, in this case there exists an index $\ell\in\bar{A}$ such that $-\ell\notin A$. Let us carry out the multiplications in (5.2). Because of the independence properties of random spectral measures each product in this expression can be factorized, and one of the factors in either $E\, Z_G(\Delta_\ell)=0$ or $E\, Z_G(\Delta_\ell)Z_G(\pm\Delta_\ell)Z_G(\pm\Delta_\ell)=0$ if $\ell\notin A$, and it is $E\, Z_G(\Delta_\ell)=0$ if $\ell\in A$.

Let

$$F=\{j_p,\ k_r;\ (p,r)\in B\} \quad \text{and} \quad \bar{F}=\{j_{\bar{p}},\ k_{\bar{r}}\ ;\ (\bar{p},\bar{r})\in B\}\ .$$

A factorization argument shows again that the expression in (5.2) equals zero if the sets $F\cup(-F)$ and $\bar{F}\cup(-\bar{F})$ are disjoint. (We can restrict ourselves to the case $A=-\bar{A}$, and in this case A is disjoint both of $F\cup(-F)$ and $(\bar{F}\cup(-\bar{F})$.) Moreover if $F\cup(-F)$ and $\bar{F}\cup(-\bar{F})$ are not disjoint then the expression on (5.2) can be estimated from above by

(5.3) $$C\varepsilon \Pi G(\Delta_{j_p})\, G(\Delta_{k_r})\, G(\Delta_{j_{\bar{p}}})\, G(\Delta_{k_{\bar{r}}}) \,,$$

with an absolute constant $C>0$, where the indices $j_p, k_r, j_{\bar{p}}, k_{\bar{r}}$ are the same as in (5.2) with the following difference: All indices appear in (5.3) with multiplicity 1, and if both indices ℓ and $-\ell$ are present in (5.2) then one of them is omitted from (5.3). The multiplying term ε appears in (5.3), since by carrying out the multiplications in (5.2) and factorizing each term, we get that all non-zero terms have a factor $E\, Z_G(\Delta)^2 Z_G(-\Delta)^2 = 3G(\Delta)^2$ or $(E|Z_G(\Delta)|^2)^2 = G(\Delta)^2$, and $G(\Delta)<\varepsilon$, $\Delta\in\mathcal{D}$

Let us estimate each term in the sum by expressing $E(\Sigma_2^{\gamma})^2$ in the above way. All terms of the form (5.3) can appear only with a multiplicity less than $C(n,m)$, where $C(n,m)$ is an appropiate constant. Hence we can write

$$E(\Sigma_2^{\gamma})^2 \le K_1^2\, K_2^2\, C(n,m)C\varepsilon \sum_{r=1}^{n+m} \sum_{j_1,\dots,j_r}'' G(\Delta_{j_1})\dots G(\Delta_{j_r}) \,,$$

where the indices $j_1,\dots,j_r \in \{\pm 1,\dots,\pm N\}$ in the sum Σ'' are all different. Hence

$$E(\Sigma_2^{\gamma})^2 \le C_1 \varepsilon \sum_{r=1}^{n+m} G(A)^r \le C_2 \varepsilon$$

with some appropriate constants C_1 and C_2. Because of (5.1), (5.1)' and the last relation one has

$$E[I_G(h_1)I_G(h_2) - \sum_{\gamma\in\Gamma} \frac{(n+m-2|\gamma|)!}{n!m!} I_G(h_\gamma)]^2 \le$$

$$\le C_3[\sum_{\gamma\in\Gamma} E((m+n-2|\gamma|)!\, I_G(h_\gamma)-\Sigma_1^\gamma)^2 + E(\Sigma_2^\gamma)^2] \le C_4\varepsilon .$$

Since $\varepsilon>0$ can be chosen arbitrarily small, part B is proved in the special case $h_1\in\hat{\bar{H}}_G^n$, $h_2\in\hat{\bar{H}}_G^m$.

If $h_1\in\bar{H}_G^n$, $h_2\in\bar{H}_G^m$ then let us choose a sequence of functions $h_{1,r}\in\hat{\bar{H}}_G^n$ $h_{2,r}\in\hat{\bar{H}}_G^m$, $r=1,2,\dots,$ such that $h_{1,r}\to h_1$ and $h_{2,r}\to h_2$ as $r\to\infty$ in the norm of the spaces $\bar{H}_G^n$ and $\bar{H}_G^m$, respectively. Define the functions $\hat{h}_\gamma(r)$ and $h_\gamma(r)$ in the same way as h_γ, but substitute the pair of functions (h_1,h_2) by $(h_{1,r}, h_2)$ and $(h_{1,r}, h_{2,r})$ in the definition. We shall show by the help of part A) that

$$E|I_G(h_1)I_G(h_2) - I_G(h_{1,r})I_G(h_{2,r})| \to 0$$

and

$$E|I_G(h_\gamma) - I_G(h_\gamma(r))| \to 0$$

as $r\to\infty$ for all $\gamma\in\Gamma$. Then a simple limiting procedure shows that Theorem 5.3 holds for all $h_1\in\bar{H}_G^n$, $h_2\in\bar{H}_G^m$.

$$E|I_G(h_1)I_G(h_2) - I_G(h_{1,r})I_G(h_{2,r})| \le$$

$$\le E|I_G(h_1-h_{1,r})I_G(h_2)| + E|I_G(h_{1,r})I_G(h_2-h_{2,r})| \le$$

$$\leq \frac{1}{n!m!}[\| h_1-h_{1,r}\|^{1/2}\| h_2\|^{1/2} + \| h_{1,r}\|^{1/2}\| h_2-h_{2,r}\|^{1/2}] \to 0,$$

and

$$E|I_G(h_\gamma)-I_G(h_\gamma(r))| \leq E|I_G(h_\gamma-\hat{h}_\gamma(r))| + E|I_G(h_\gamma(r)-\hat{h}_\gamma(r))| \leq$$

$$\leq \| h_\gamma-\hat{\hat{h}}_\gamma(r)\|^{1/2} + \| h_\gamma(r)-\hat{h}_\gamma(r)\|^{1/2} \leq$$

$$\leq \| h_1-h_{1,r}\|^{1/2}\| h_2\|^{1/2} + \| h_{1,r}\|^{1/2}\| h_2-h_{2,r}\|^{1/2} \to 0$$

Theorem 5.3 is proved.

Now we formulate some consequences of Theorem 5.3. Let $\bar{\Gamma}\subset\Gamma$ denote the set of complete diagrams, i.e. let a diagram $\gamma\in\bar{\Gamma}$ if an edge enters into each vertex of γ. $E\,I(h_\gamma) = 0$ for all $\gamma\in\Gamma-\bar{\Gamma}$, since (4.3) holds for all $f\in\bar{H}_G^n$, $n\geq 1$. If $\gamma\in\bar{\Gamma}$, then $I(h_\gamma)\in H_G^0$. Let h_γ denote the value of $I(h_\gamma)$ in this case.

Now we have the following

Corollary 5.4

$$E\, I_G(h_1)\ldots I_G(h_m) = \sum_{\gamma\in\bar{\Gamma}} \frac{1}{n_1!\ldots n_m!}\, h_\gamma$$

(The sum on the right-hand side equals zero if $\bar{\Gamma}$ is empty.)

In the next result we get a formula for the expectation of products of Hermite polynomials.

Corollary 5.5

Let $(X_1,\dots,X_p)$, $p\geq 2$, be a Gaussian random vector $E X_j=0$, $EX_j^2=1$, $E X_jX_k=r_{j,k}$, $j,k=1,2,\dots,p$. Then

$$E H_{k_1}(X_1)\dots H_{k_p}(X_p)=\begin{cases}\dfrac{k_1!\dots k_p!}{2^q q!}\sum{}' r_{i_1,j_1}\dots r_{i_q,j_q} \\ \quad \text{if } k_1+\dots+k_p=2q,\ 0\leq k_1,\dots,k_q\leq q \\ 0 \quad \text{otherwise}\end{cases}$$

where $\sum{}'$ is a sum over all indices $i_1,j_1,\dots,i_q,j_q$

(i) $i_1,j_1,\dots,i_q,j_q\in\{1,2,\dots,p\}$

(ii) $i_1\neq j_1,\dots,i_q\neq j_q$

(iii) there are k_1 indices 1, k_2 indices $2,\dots,k_p$ indices p.

Proof of Corollary 5.5

We can represent the random variables X_j in the form $X_j=\sum_k c_{j,k}\,\xi_k$ with some appropiate coefficients $c_{j,k}$, where $\xi_1,\xi_2,\dots$ is a sequence of independent standard normal random variables. Let $Z(dx)$ denote a random spectral measure corresponding to the one-dimensional spectral measure with density function $g(x)=\frac{1}{2\Pi}$ for $|x|<\Pi$, $g(x)=0$ for $|x|>\Pi$. The standard normal random variables $\int \exp(ikx)\,Z(dx)$, $k=1,2,\dots$ are independent. Define $h_j(x)=\sum_k c_{j,k}\exp(i k x)$, $j=1,2,\dots,p$. The random variables X_j can be identified with

$\int h_j(x)Z(dx)$, $j=1,\dots,p$, since the corresponding joint distributions coincide. Let $\bar{h}_j(x_1,\dots,x_{k_j}) = \prod_{\ell=1}^{k_j} h_j(x_\ell)$.

By Itô's formula $H_{k_j}(X_j) = \int \bar{h}_j(x_1,\dots,x_{j_k})Z(dx_1)\dots Z(dx_{j_k})$.

An application of Corollary 5.4 for $\bar{h}_{j_1},\dots,\bar{h}_{j_k}$ yields Corollary 5.5. To see this, one has to observe that

$\int_{-\pi}^{\pi} h_j(x)\overline{h_k(x)}dx = 0$, $j\neq k$, hence the same products

$r_{i_1,j_1}\dots r_{i_q,j_q}$ appear in this application of Corollary 5.4 as in Corollary 5.5. If the term $r_{i,j}$ appears with multiplicity $a(i,j)$ in such a product, then this product was counted $k_1!\dots k_p![\prod a(j,k)!]^{-1}$ times in Corollary 5.4 with a coefficient $(k_1!\dots k_m!)^{-1}$ and $2^q q!\,[\prod a(j,k)!]^{-1}$ times in Corollary 5.5 with a coefficient $(2^q q!)^{-1}$.

Theorem 5.3 states, in particular, that the product of Wiener-Itô integrals with respect to the random spectral measure of a stationary Gaussian field belongs to the Hilbert space $\mathcal{H}$ defined by this field, since it can be written as a sum of Wiener-Itô integrals. This means a trivial measurability condition, and also that the product has a finite second moment, what is not so trivial. Theorem 5.3 actually gives the following non-trivial inequality.

Let $h_1 \in \mathcal{H}_G^{n_1},\dots,h_m \in \mathcal{H}_G^{n_m}$. Let $C(n_1,\dots,n_m) =$

$= \dfrac{|\bar{\Gamma}(n_1,n_1,\dots,n_m,n_m)|}{n_1!\dots n_m!}$ where $|\bar{\Gamma}(n_1,n_1,\dots,n_m,n_m)|$

denotes the cardinality of complete diagrams in

$\Gamma(n_1,n_1,\ldots,n_m, n_m)$. In the special case $n_1 = \ldots = n_m = n$ let $\overline{C}(n,m) = C(n_1,\ldots,n_m)$. Then

Corollary 5.6

$$E[I_G(h_1)^2 \ldots I_G(h_m)^2] \le C(n_1,\ldots,n_m) EI_G(h_1)^2 \ldots EI_G(h_m^2).$$

In particular

$$E[I_G(h)^{2m}] \le \overline{C}(n,m)(E(I_G(h)^2))^m$$

if $h \in H_G^n$.

Corollary 5.6 follows immediately from Corollary 5.4 with the choice $h_1,h_1,\ldots,h_m,h_m$. One only has to observe that $|h_\gamma| \le ||h_1||^2 \ldots ||h_m||^2$ for all complete diagrams by part A of Theorem 5.3. The inequality in Corollary 5.6 is sharp. If G is a finite measure and $h_1 \in H_G^{n_1},\ldots,h_m \in H_G^{n_m}$ are constant functions, then equality can be written in Corollary 5.6. We remark that in case $I_G(h_1),\ldots,I_G(h_m)$ are constant times the n_1-th ..., n_m-th Hermite polynomials of the same standard normal random variable. Let us emphasize that the constant $C(n_1,\ldots,n_m)$ depends only on the parameters $n_1,\ldots,n_m$, and not on the form of $h_1,\ldots,h_m$. The coefficient $C(n_1,\ldots,n_m)$ is a monotone function of its arguments. The following argument shows that

$$C(n_1+1,n_2,\ldots,n_m) \ge C(n_1,n_2,\ldots,n_m).$$

Let us say that two complete diagrams in $\overline{\Gamma}(n_1,n_1,\ldots,n_m,n_m)$ or in $\overline{\Gamma}(n_1+1,n_1+1,\ldots,n_m,n_m)$ are equivalent if they can be transformed into each other by permuting the vertices $(1,1),\ldots;(1,n_1)$ in $\overline{\Gamma}(n_1,n_1,\ldots,n_m,n_m)$ or the vertices

$(1,1),\ldots,(1,n_1+1)$ in $\bar{\Gamma}(n_1+1,n_1+1,\ldots,n_m,n_m)$. The equivalence classes have $n_1!$ elements in the first case and $(n_1+1)!$ elements in the second one. Moreover the number of equivalence classes is less in the first case than in the second one. (They would agree if we counted only those equivalence classes in the second case which contain a diagram where $(1,n_1+1)$ and $(2,n_1+1)$ are connected by an edge.) Hence

$$\frac{1}{n_1!}|\bar{\Gamma}(n_1,n_1,\ldots,n_m,n_m)| \le \frac{1}{(n_1+1)!}|\bar{\Gamma}(n_1+1,n_1+1,\ldots,n_m,n_m)|$$

as we claimed.

The next result, formulated in a more elementary way, may better illuminate the content of Corollary 5.6.

Corollary 5.7

Let $\xi_1,\ldots,\xi_k$ be a normal random vector, and $P(x_1,\ldots,x_k)$ a polynomial of degree n. Then

$$E[P(\xi_1,\ldots,\xi_k)^{2m}] \le \bar{C}(n,m)(n+1)^m(E(P(\xi_1,\ldots,\xi_k)^2))^m$$

The multiplying constant $\bar{C}(n,m)(n+1)^m$ is not sharp in this inequality. Observe that it does not depend on k.

Proof of Corollary 5.7

We can write $\xi_j = \int f_j(x)Z(dx)$ with some $f_j \in H^1$, $j=1,2,\ldots,k$, where $Z(dx)$ is the same as in the proof of Corollary 5.5. There exist some $h_j \in H^j$, $j=0,1,\ldots,n$, such that

$$P(\xi_1,\dots,\xi_k) = \sum_{j=o}^{n} I(h_j) .$$

Then

$$E[P(\xi_1,\dots,\xi_k)^{2m}] = E[(\sum_{j=o}^{n} I(h_j))^{2m}] \le (n+1)^m E((\sum_{j=o}^{n} I(h_j)^2)^m) \le$$

$$\le (n+1)^m \sum_{p_o+\dots+p_n=m} C(p_1,\dots,p_n)[EI(h_o)^2]^{p_o}\dots[EI(h_n)^2]^{p_n} \frac{m!}{p_1!\dots p_n!} \le$$

$$\le (n+1)^m \bar{C}(n,m) \sum_{p_o+\dots+p_n=m} [EI(h_o)^2]^{p_o}\dots[EI(h_n)^2]^{p_n} \frac{m!}{p_1!\dots p_n!} =$$

$$= (n+1)^m \bar{C}(n,m) E[\sum I(h_j)^2]^m \le (n+1)^m C(n,m)[EP(\xi_1,\dots,\xi_k)^2]^m .$$

6. Subordinated fields, Construction of self-similar fields

Let X_n, $n \in Z_\nu$, be a discrete stationary Gaussian field, and let the field ξ_n, $n \in Z_\nu$, be subordinated to it. Let Z_G denote the random spectral measure adapted to the field X_n. By Theorem 4.1 ξ_o can be represented in the form

$$\xi_o = f_o + \sum_{n=1}^{\infty} \frac{1}{n!} \int f_n(x_1,\dots,x_n) Z_G(dx_1)\dots Z_G(dx_n) .$$

$f = (f_o, f_1, \dots) \in \mathrm{Exp} H_G$ in a unique way. Theorem 4.3 yields the following

Theorem 6.1

The field ξ_n, $n \in Z_\nu$, subordinated to the stationary Gaussian field X_n, $n \in Z_\nu$, can be written in the form

$$(6.1)\quad \xi_j = f_o + \sum_{n=1}^{\infty} \frac{1}{n!} \int \exp[i(j, x_1+\dots+x_n)] f_n(x_1,\dots,x_n) Z_G(dx_1) \dots Z_G(dx_n)\,, \quad j \in Z_\nu\,.$$

with some $f = (f_o, f_1, \dots) \in \mathrm{Exp}H_G$, where G is the spectral measure of the field X_n, and Z_G is the random spectral measure adapted to it. This representation is unique.

If the spectral measure G is non-atomic, then the functions $\bar{f}_n(x_1,\dots,x_n) = f_n(x_1,\dots,x_n)\,\tilde{\chi}_o^{-1}(x_1+\dots+x_n)$ are meaningful, where $\tilde{\chi}_j(x) = \exp[i(j,x)] \prod_{k=1}^{\nu} \frac{\exp(ix^{(k)})-1}{ix^{(k)}}$, $j \in Z_\nu$, is the Fourier transform of the uniform distribution on the unit cube $\times_{k=1}^{\nu} [j^{(k)}, j^{(k)}+1)$. Then ξ_j can be rewritten as

$$\xi_j = f_o + \sum_{n=1}^{\infty} \frac{1}{n!} \int \tilde{\chi}_j(x_1+\dots+x_n) \bar{f}_n(x_1,\dots,x_n) Z_G(dx_1)\dots Z_G(dx_n).$$

Hence the following Theorem 6.1' can be considered as the continuous time version of Theorem 6.1.

Theorem 6.1'

Let the generalized field $\xi(\varphi)$, $\varphi \in S$, be subordinated to the stationary Gaussian generalized field $X(\varphi)$, $\varphi \in S$. Let G denote the spectral measure of the field X, and let Z_G be the random spectral measure adapted to it. Then $\xi(\varphi)$ can be written in the form

$$(6.1')\quad \xi(\varphi) = f_o \tilde{\varphi}(0) + \sum_{n=1}^{\infty} \frac{1}{n!} \int \tilde{\varphi}(x_1+\dots+x_n) f_n(x_1,\dots,x_n) Z_G(dx_1) \dots Z_G(dx_n)\,,$$

where f_n is invariant under permutations of its variables, $f_n(-x_1,\ldots,-x_n) = \overline{f_n(x_1,\ldots,x_n)}$, $n=1,2,\ldots$ and

(6.2) $\sum \frac{1}{n!} \int [1+|x_1+\ldots+x_n|]^q |f_n(x_1,\ldots,x_n)|^2 G(dx_1)\ldots G(dx_n) < \infty$

with an appropriate $q > 0$. This representation is unique.

Proof of Theorem 6.1'

The relation

(6.3) $\xi(\varphi)=\Phi_{\varphi,0}+\sum_{n=1}^{\infty}\frac{1}{n!}\int \Phi_{\varphi,n}(x_1,\ldots,x_n)Z_G(dx_1)\ldots Z_G(dx_n)$

with $(\Phi_{\varphi,0},\Phi_{\varphi,1},\ldots) \in \operatorname{Exp} H_G$ holds true. We are going to show that

$$\Phi_{\varphi,n}(x_1,\ldots,x_n)=f_n(x_1,\ldots,x_n)\tilde{\varphi}(x_1+\ldots+x_n), \quad n=1,2,\ldots$$

and

$$\Phi_{\varphi,0} = f_0\tilde{\varphi}(0) \quad \text{for all} \quad \varphi \in S .$$

Let us choose a $\varphi_0 \in S$ such that $\varphi_0(x) > 0$ for all $x \in R^\nu$. We claim that the finite linear combinations $\sum a_k\varphi_0(x+t_k)$ are dense in S. It is enough to show that all $\psi \in S$ whose Fourier transform $\tilde{\psi}$ has a compact support, can arbitrary well be approximated by such linear combinations, because these functions ψ are dense in S. The function $\frac{\tilde{\psi}}{\tilde{\varphi}_0} \in S^c$, ($S^c$ denotes the Schwartz-space of complex valued functions again), because $\tilde{\varphi}_0(x) \neq 0$ and $\tilde{\psi}$ has a compact support. There exists a $\chi \in S$ such that $\tilde{\chi} = \frac{\tilde{\psi}}{\tilde{\varphi}_0}$. Therefore $\psi(x) = \int\chi(-t)\varphi_0(x+t)dt$. It is not difficult to see, exploiting this relation together with the rapid decrease of χ at infinity, that for all $r > 0$, $s > 0$ and $\varepsilon > 0$ there exists a linear combination $\hat{\psi}(x) = \sum a_k \varphi_0(x+t_k)$ such that $(1+|x|^s)|\psi(x)-\hat{\psi}(x)| < \varepsilon$, and the same relation holds for the derivatives of $\psi(x) - \hat{\psi}(x)$ of order less than r. This fact implies that the linear combinations $\sum a_k \varphi_0(x+t_k)$ are dense in S.

Define

$$f_n(x_1,\dots,x_n) = \frac{\Phi_{\varphi_0}(x_1+\dots+x_n)}{\tilde{\varphi}_0(x_1+\dots+x_n)}, \quad n=1,2,\dots, \quad f_0 = \Phi_{\varphi_0}[\tilde{\varphi}(0)]^{-1}.$$

If $\varphi(x) = \sum a_k \varphi(x+t_k)$, then

$$\xi(\varphi) = f_0(\sum a_k)\tilde{\varphi}_0(0) + \sum_{n=1}^{\infty} \int [\sum a_k e^{i(t_k, x_1+\dots+x_n)} \tilde{\varphi}_0(x_1+\dots+x_n)]$$

$$f_n(x_1\dots,x_n) \; Z_G(dx_1)\dots Z_G(dx_n) =$$

$$= f_0 \; \tilde{\varphi}(0) + \sum_{n=1}^{\infty} \int \tilde{\varphi}(x_1+\dots x_n) f_n(x_1,\dots,x_n) Z_G(dx_1)\dots Z_G(dx_n)$$

Relation (6.3) holds for an arbitrary $\varphi \in S$. Moreover there exists a sequence $\varphi_j \in S$ satisfying (6.1)' such that $\varphi_j \to \varphi$ in the topology of S. As $\lim E[X(\varphi_j)-X(\varphi)]^2 = 0$

$$\int_A |\Phi_{\varphi,n}(x_1,\dots,x_n) - \tilde{\varphi}_j(x_1+\dots+x_n) f_n(x_1,\dots,x_n)|^2 G(dx_1)\dots G(dx_n) \to 0$$

as $j \to \infty$ for all n and for all compact sets $A \subset R^{n\nu}$. On the other hand

$$\int_A |\tilde{\varphi}(x_1+\ldots+x_n)-\tilde{\varphi}_j(x_1+\ldots+x_n)|^2 |f_n(x_1,\ldots,x_n)|^2 G(dx_1)$$

$$\ldots G(dx_n) \to 0 \quad ,$$

since $\tilde{\varphi}_j \to \tilde{\varphi}$ in the topology of S . Therefore

$$\Phi_{\varphi,n}(x_1,\ldots,x_n)=\tilde{\varphi}(x_1+\ldots+x_n)f_n(x_1,\ldots,x_n), \; n=1,2,\ldots$$

Similarly

$$\Phi_{\varphi,0} = \tilde{\varphi}(0)f_0 \; .$$

These relations imply (6.1)'. To complete the proof of Theorem 6.1 we show that (6.2) is implied by the continuity of the transformation $F: \varphi \to X(\varphi)$ from the space S into the space $L^2(\Omega,A,P)$.

We recall that the transformation $\varphi \to \tilde{\varphi}$ is bicontinuous in S^c . Therefore the continuity of the transformation F implies that for all $\varepsilon > 0$ there exist $p > 0$, $r > 0$ and $\delta > 0$ such that if

$$(6.4) \quad (1+|x|^p)\left|\frac{\partial^{s_1+\ldots+s_\nu}}{\partial x^{(1)^{s_1}}\ldots\partial x^{(\nu)^{s_\nu}}}\tilde{\varphi}(x)\right|<\delta \quad \text{for all}$$

$$s_1+\ldots+s_\nu \leq r$$

then $EX(\varphi)^2 < \varepsilon$.

Let us choose a $\psi \in S$ such that ψ has a compact support, $\psi(x) = \psi(-x)$, $\psi(x) \geq 0$ for all $x \in R^\nu$, and $\psi(x) = 1$ if $|x| \leq 1$. Define the functions $\tilde{\varphi}_m(x) = C(1+|x|)^{-p} \psi(\frac{x}{m})$. Then $\varphi_m \in S$, and φ_m satisfies (6.4) if $C > 0$ is sufficiently small. Hence

$$EX(\varphi_m)^2 = \sum \frac{1}{n!} \int |\tilde{\varphi}_m(x_1+\dots+x_n)|^2 |f_n(x_1,\dots,x_n)|^2 G(dx_1)\dots G(dx_n) < \varepsilon$$

for all $m = 1,2,\dots$

As $\tilde{\varphi}_m(x) \to C(1+|x|)^{-p}$ as $m \to \infty$, and $\tilde{\varphi}_m(x) \geq 0$, the last inequality together with Fatou's lemma imply (6.2). Theorem 6.1 is proved.

We shall call the representations given in Theorems 6.1 and 6.1' the canonical representation of a subordinated field. From now on we restrict ourselves to the case $E\xi_n = 0$ or $E\xi(\varphi) = 0$ respectively, i.e. to the case when $f_0 = 0$ in the canonical representation. If

$$\xi(\varphi) = \sum_{n=1}^{\infty} \frac{1}{n!} \int \tilde{\varphi}(x_1+\dots+x_n) f_n(x_1,\dots,x_n) Z_G(dx_1)\dots Z_G(dx_n)$$

then

$$\xi(\varphi_t^A) = \sum_{n=1}^{\infty} \frac{1}{n!} \frac{t^\nu}{A(t)} \int \tilde{\varphi}\left(\frac{x_1+\dots+x_n}{t}\right) f_n(x_1,\dots,x_n) Z_G(dx_1)\dots Z_G(dx_n)$$

where φ_t^A is defined in (1.3). Define the spectral measures G_t by the formula $G_t(A) = G(tA)$. Then

$$\xi(\varphi_t^A) \stackrel{\Delta}{=} \sum_{n=1}^{\infty} \frac{1}{n!} \frac{t^\nu}{A(t)} \int \tilde{\varphi}(x_1+\dots+x_n) f_n\left(\frac{x_1}{t},\dots,\frac{x_n}{t}\right) Z_{G_t}(dx_1)\dots Z_{G_t}(dx_n) .$$

If $G(Bt) = t^{2\varkappa}G(B)$ with some $\varkappa > 0$ for all $t > 0$ and $B \in \mathcal{B}^\nu$, $f_n(\lambda x_1,\dots,\lambda x_n) = \lambda^{\nu-\varkappa n-\alpha} f(x_1,\dots,x_n)$, and $A(t)$ is chosen as $A(t) = t^\alpha$, then Theorem 4.4 implies that $\xi(\varphi_t^A) \overset{\Delta}{=} \xi(\varphi)$.

Hence we obtain the following:

Theorem 6.2

Let the generalized field $\xi(\varphi)$ be given by the formula

$$(6.5)\quad \xi(\varphi) = \sum_{n=1}^{\infty} \frac{1}{n!} \int f_n(x_1,\dots,x_n)\tilde{\varphi}(x_1+\dots+x_n)Z_G(dx_1\dots Z_G(dx_n)$$

If $f_n(\lambda x_1,\dots,\lambda x_n) = \lambda^{\nu-\varkappa n-\alpha} f(x_1,\dots,x_n)$ for all n, $(x_1,\dots,x_n) \in R^{\nu n}$ and $\lambda > 0$, $G(\lambda A) = \lambda^{2\varkappa}G(A)$ for all $\lambda > 0$ and $A \in \mathcal{B}^\nu$, then ξ is a self-similar field with self-similarity parameter α.

The discrete time version of this result can be proved in the same way. It states the following

Theorem 6.2'

If the discrete field ξ_n, $n \in Z_\nu$, has the form

$$(6.5')\quad \xi_j = \sum_{n=1}^{\infty} \frac{1}{n!} \int f_n(x_1,\dots,x_n)\tilde{\chi}_j(x_1+\dots+x_n)Z_G(dx_1)\dots Z_G(dx_n),$$

$$j \in Z_\nu$$

and $f_n(\lambda x_1,\dots,\lambda x_n) = \lambda^{\nu-\varkappa n-\alpha} f_n(x_1,\dots,x_n)$, $G(\lambda A) = \lambda^{2\varkappa}G(A)$

then ξ_n is a self-similar field with self-similarity parameter α.

Theorems 6.2 and 6.2' enable us to construct self-similar fields. Nevertheless, we have to check whether formulae (6.5) and (6.5') are meaningful. The hard part of this problem is to check whether

$$\sum \frac{1}{n!} \int |f_n(x_1,\dots,x_n)|^2 |\tilde{\chi}_j(x_1+\dots+x_n)|^2 G(dx_1)\dots G(dx_n) < \infty$$

or whether

$$\sum \frac{1}{n!} \int |f_n(x_1,\dots,x_n)|^2 |\tilde{\varphi}(x_1+\dots+x_n)|^2 G(dx_1)\dots G(dx_n) < \infty$$

for all $\varphi \in S$.

To investigate when these expressions are finite is a rather hard problem in the general case. The next result enables us to prove the finiteness of these expressions in some interesting cases.

Let us define the measure G

(6.6) $\quad G(A) = \int_A |x|^{2\varkappa-\nu} \; a(\frac{x}{|x|})dx, \quad A \in \mathcal{B}^\nu$,

where a (.) is a non-negative measurable function on the ν-dimensional unit sphere $S_{\nu-1}$. We prove the following

Proposition 6.3

Let the measure G be the same as in formula (6.6).

a) If the function a (.) is bounded on the unit sphere $S_{\nu-1}$

and $\frac{\nu}{n} > 2\varkappa > 0$, then

$$D(\varphi) = \int |\tilde{\varphi}(x_1+\ldots+x_n)|^2 G(dx_1)\ldots G(dx_n) < \infty \quad \text{for all} \quad \varphi \in S$$

and

$$D(j) = \int |\tilde{\chi}_j(x_1+\ldots+x_n)|^2 G(dx_1)\ldots G(dx_n) < \infty \quad \text{for all} \quad j \in Z_\nu .$$

b) If there is a constant $C > 0$ such that $a(x) > C$ in a neighbourhood of a point $x_0 \in S_{\nu-1}$ and either $2\varkappa \leq 0$ or $2\varkappa \geq \frac{\nu}{n}$ then the integrals $D(j)$ and some $D(\varphi)$, $\varphi \in S$, are divergent.

Proof of Proposition 6.3

Proof of part a) We may assume that $a(x) \equiv 1$, $x \in S_{\nu-i}$.

Define

$$J_{\varkappa,n}(x) = \int_{x_1+\ldots+x_n=x} |x_1|^{2\varkappa-\nu} \ldots |x_n|^{2\varkappa-\nu} dx_1\ldots dx_n , \quad x \in R^\nu$$

where $dx_1\ldots dx_n$ denotes the Lebesgue measure on the hyperplane $x_1+\ldots+x_n = x$.

$$J_{\varkappa,n}(\lambda x) = |\lambda|^{n(2\varkappa-\nu)+(n-1)\nu} J_{\varkappa,n}(x), \quad x \in R^\nu, \quad \lambda \in R^1$$

because of the homogeneity of the integral, and

$$(6.7) \qquad D(j) = \int_{R^\nu} |\tilde{\chi}_j(x)|^2 J_{\varkappa,n}(x)dx ,$$

$$D(\varphi) = \int |\tilde{\varphi}(x)|^2 J_{\varkappa,n}(x)dx$$

We prove, by induction on n, that

$$(6.8) \qquad J_{\varkappa,n}(x) \le C(\varkappa,n)|x|^{2\varkappa n-\nu}$$

with an appropriate $C(\varkappa,n) < \infty$ if $\frac{\nu}{n} > 2\varkappa > 0$.

$$J_{\varkappa,n}(x) = \int J_{\varkappa,n-1}(y)J_{\varkappa,1}(x-y)dy$$

hence

$$J_{\varkappa,n}(x) \le C(\varkappa,n-1)C(\varkappa,1)\int |y|^{2\varkappa(n-1)-\nu}|x-y|^{2\varkappa-\nu}dy =$$

$$= C(\varkappa,n-1)C(\varkappa,1)|x|^{2\varkappa n-\nu} \int |y|^{2\varkappa(n-1)-\nu}\left|\frac{x}{|x|} -y\right|^{2\varkappa-\nu}dy =$$

$$\doteq C(\varkappa,n)|x|^{2\varkappa n-\nu},$$

since $\int |y|^{2\varkappa(n-1)-\nu}\left|\frac{x}{|x|} -y\right|^{2\varkappa-\nu}dx < \infty$.

The last integral is finite since its integrand behaves at zero asymptotically as $c_1|y|^{2\varkappa(n-1)-\nu}$ at $e = \frac{x}{|x|} \in S_{\nu-1}$ as $c_2|y-e|^{2\varkappa-\nu}$ and at infinity as $c_3|y|^{2\varkappa n-2\nu}$.

Relations (6.7) and (6.8) imply that

$$D(j) \le C'\int |\tilde{\chi}(x)|^2|x|^{2\varkappa n-\nu}dx \le C''\int |x|^{2\varkappa n-\nu}\prod_{\ell=1}^{\nu}\frac{1}{1+|x^{(\ell)}|^2}dx =$$

$$= C''\nu\int_{|x^{(1)}|=\max_{1\le\ell\le\nu}|x^{(\ell)}|} = C''\nu\Big[\sum_{k=o}^{\infty}\int_{\substack{|x^{(1)}|=\max|x^{(\ell)}| \\ 2^k\le|x^{(1)}|<2^{k+1}}} +$$

$$+ \int_{\substack{|x^{(1)}|=\max|x^{(\ell)}| \\ |x^{(1)}|<1}}\Big].$$

Hence

$$D(j) \le C_1 \sum_{k=0}^{\infty} 2^{k(2\varkappa n-\nu)} [\int_{-\infty}^{\infty} \frac{1}{1+x^2} dx]^{\nu} + C_2 < \infty \quad ,$$

The proof of the relation $D(\varphi) < \infty$ is similar but simpler.

Proof of part b)

Since

$$J_{\varkappa,n}(x) \ge \int_{\substack{|y|<(\frac{1}{2}+\alpha)|x| \\ |y-x|<(\frac{1}{2}+\alpha)|x|}} J_{\varkappa,n-1}(y) J_{\varkappa,1}(x-y)\,dy$$

with an arbitrary $\alpha>0$, an argument similar to the one in part a) shows that

$$J_{\varkappa,n,a} \begin{cases} \ge \bar{C}(\varkappa,n)|x|^{2\varkappa n-\nu} & \text{if} \quad \frac{\nu}{n} > 2\varkappa > 0 \\ = \infty & \text{if} \quad \varkappa \le 0 \quad \text{or} \quad 2\varkappa \ge \frac{\nu}{n} \end{cases}$$

in a neighbourhood of e. Since $|\tilde{\chi}_j(x)|^2 > 0$ for almost all $x \in R^{\nu}$,

$$D(j) = \int |\tilde{\chi}_j(x)|^2 J_{\varkappa,n,a}(x)\,dx = \infty$$

under the conditions of part b). Similarly $D(\varphi) = \infty$ if $|\tilde{\varphi}(x)|^2 > 0$ for almost all $x \in R^{\nu}$. We remark that the conditions in part b) can be weakened. It would have been enough to assume that $a(x) > 0$ on a set of positive Lebesgue measure in $S_{\nu-1}$.

Theorems 6.2 and 6.2' together with Proposition 6.3 have the following

Corollary 6.4

The formulae

$$X_j = \sum_{n=1}^{M} \int \tilde{\chi}_j(x_1+\dots+x_n) \prod_{\ell=1}^{n} (|x_\ell|^{-\varkappa+\frac{\nu-\alpha}{n}} b_n(\frac{x_\ell}{|x_\ell|})) Z_G(dx_1)\dots Z_G(dx_n)$$

$j \in Z_\nu$

and

$$X(\varphi) = \sum_{n=1}^{M} \int \tilde{\varphi}(x_1+\dots+x_n) \prod_{\ell=1}^{n} (|x_\ell|^{-\varkappa+\frac{\nu-\alpha}{n}} b_n(\frac{x_\ell}{|x_\ell|})) Z_G(dx_1)\dots$$

$Z_G(dx_n)$, $\varphi \in S$

define self-similar fields with self-similarity parameter α if G is defined by formula (6.6), the parameter α satisfies the inequality $\frac{\nu}{2} < \alpha < \nu$, and $a(.)$, $b_1(.)$ ..., $b_n(.)$ are bounded even functions on $S_{\nu-1}$.

Remark 6.5

The estimate on $J_{\varkappa,n}$ and the end of the proof of part a) in Proposition 6.3 show that the self-similar fields

$$X(\varphi) = \int \tilde{\varphi}(x_1+\dots+x_n) |x_1+\dots+x_n|^p u(\frac{x_1+\dots+x_n}{|x_1+\dots+x_n|}) \prod_{\ell=1}^{n} (|x_\ell|^{-\varkappa+\frac{\nu-\alpha}{n}}$$

$$b(\frac{x_\ell}{|x_\ell|}))\ Z_G(dx_1)\dots Z_G(dx_n), \quad \varphi \in S$$

and

$$X(j)=\int \tilde{\chi}_j(x_1+\dots+x_n)\,|x_1+\dots+x_n|^p u\left(\frac{x_1+\dots+x_n}{|x_1+\dots+x_n|}\right) \prod_{\ell=1}^{n} \left(|x_\ell|^{-\varkappa+\frac{\nu-\alpha}{n}}\right.$$

$$\left. b\left(\frac{x_\ell}{|x_\ell|}\right)\right) Z_G(dx_1)\dots Z_G(dx_n)$$

are well defined if G is defined by formula (6.6), $a(.)$, $b(.)$ and $u(.)$ are bounded even functions on $S_{\nu-1}$, $\frac{\nu}{2} < \alpha < \nu$, and $\alpha-p<\nu$ in the generalized and $\frac{\nu-1}{2} < \alpha-p < \nu$ in the discrete field case. The self-similarity parameter of these fields is $\alpha-p$. We remark that in the case $p > 0$ this class of self-similar fields also contains self-similar fields with self-similarity parameter less than $\nu/2$.

The following question arises in a natural way: When do different formulae satisfying the conditions of Theorem 6.2 or Theorem 6.2' define self-similar fields with different distributions? In particular: are the self-similar fields constructed via multiple Wiener-Itô integrals necessarily non-Gaussian? We cannot give a completely satisfactory answer for the above question, but our former results yield some useful informations. Let us substitute the spectral measure G by G' such that $\frac{G(dx)}{G'(dx)} = |g(x)|^2$, $g(-x) = \overline{g(x)}$ and the functions $|x_j|^{-\varkappa+\frac{\nu-\alpha}{n}} b\left(\frac{x_j}{|x_j|}\right)$ by

$b(\frac{x_j}{|x_j|})q(x_j)|x_j|^{-\varkappa+\frac{\nu-\alpha}{n}}$ in Corollary 6.4. By Theorem 4.4 the new field has the same distribution as the original one. On the other hand Corollary 5.4 helps us to decide whether the random variables of two random fields have different moments, and therefore the two fields have different distributions. Let us consider e.g. a moment of odd order of the random variables X_j or $X(\varphi)$ defined in Corollary 6.4. It is clear that all $h_\gamma \geq 0$. Moreover if $b_n(x)$ does not vanish for some even n then there exists a $h_\gamma > 0$ in the sum expressing an odd moment of X_j or $X(\varphi)$. Hence the odd moments of X_j or $X(\varphi)$ are positive in this case. This means in particular that the self-similar fields defined in Corollary 6.4 are non-Gaussian if b_n is non-vanishing for some even n .

The next result shows that the tail behaviour of multiple Wiener-Itô integrals of different order is different.

<u>Theorem 6.6</u>

<u>Let G be a spectral measure and Z_G a random spectral measure corresponding to G . For all $h \in H_G^m$ there exist some constants $K_1 > K_2 > 0$ and $x_0 > 0$ depending on h such that</u>

$$\exp(-K_1 x^{\frac{2}{m}}) \leq P(|I_G(h)| > x) \leq \exp(-K_2 x^{\frac{2}{m}})$$

<u>for all $x > x_0$.</u>

Proof of Theorem 6.6

a) Proof of the upper estimate

$$P(|I_G(h)|>x) \le x^{-2N} E(|I_G(h)|^{2N})$$

By Corollary 5.6

$$E(|I_G(h)|^{2N}) \le \bar{C}(m,N)(E[I_G(h)^2])^N = \bar{C}(m,N)C_1^N .$$

By a simple combinatorial argument we obtain that

$$\bar{C}(m,N) \le \frac{(2Nm-1)(2Nm-3)\dots 1}{(m!)^N}$$

since the numerator on the right hand side equals the number of complete diagrams $|\bar{\Gamma}(\underbrace{m,\dots,m}_{2N \text{ times}})|$ if vertices from the same row can also be connected. Multiplying the inequalities

$$(2Nm-2j-1)(2Nm-2j-1-2N)\dots(2Nm-2j-1-2N(m-1) \le (2N)^m \; m!$$

$j=1,\dots,N,$ we obtain that

$$\bar{C}(m,N) \le (2N)^{mN}$$

(This inequality could be sharpened, but it is sufficient for our purpose.) Choose a sufficiently small $\alpha > 0$, and define $N = [\alpha x^{\frac{2}{m}}]$, where $[\ ]$ denotes integer part. Then

$$P(|I_G(h)|>x) \le (x^{-2}(2\alpha)^m x^2)^N \; C_1^N = [C_1(2\alpha)^m]^N \le \exp(-K_1 x^{\frac{2}{m}}) ,$$

if α is chosen in such a way that $C_1(2\alpha)^m \le \frac{1}{e}$,
$K_1 = \frac{\alpha}{2}$, and $x > x_o$ for an appropriate $x_o > 0$.

b) Proof of the lower estimate

First we reduce this inequality to the following statement:
Let $Q(x_1,\dots,x_k)$ be a homogeneous polynomial of order m (k is arbitrary), and $\xi = (\xi_1,\dots,\xi_k)$ a k-dimensional standard normal vector random variable. Then

$$(6.9) \qquad P(|Q(\xi_1,\dots,\xi_k)|>x) \ge \exp(-Kx^{\frac{2}{m}})$$

if $x > x_o$, where the constants $K > 0$, $x_o > 0$ may depend on the polynomial Q .

By the results of Section 4, $I_G(h)$ can be written in the form

$$(6.10) \qquad I_G(h) = \sum_{j_1+\dots+j_\ell=m} C^{k_1\dots,k_\ell}_{j_1,\dots,j_\ell} H_{j_1}(\xi_{k_1})\dots H_{j_\ell}(\xi_{k_\ell})$$

where $\xi_1, \xi_2, \dots$ are independent standard normal random variables, the numbers C are appropriate coefficients, and the right-hand side of (6.10) is convergent in L_2 sense. Let us fix a sufficiently large integer k , and let us consider the conditional distribution of the right-hand side of (6.10) under the condition $\xi_{k+1} = x_{k+1}, \xi_{k+2} = x_{k+2},\dots,$ where the numbers $x_{k+1}, x_{k+2},\dots$ are arbitrary. This conditional distribution coincides with the distribution of the random variable $Q(\xi_1,\dots,\xi_k,x_{k+1},\dots)$ with probability 1, where the polynomial Q is obtained by substituting $\xi_{k+1} = x_{k+1}, \xi_{k+2} = x_{k+2}, \ldots$ into the right-hand side of (6.10). It is clear that all the polynomials $Q(\xi_1, \ldots, \xi_k, x_{k+1}, \ldots)$ are of order m if k is sufficiently large. It is sufficient to prove that

$$P(|Q(\xi_1, \ldots, \xi_k, x_{k+1}, \ldots)|>x) \geq \exp(-Kx^{\frac{2}{m}})$$

for $x > x_o$, where the constants $K > 0$, $x_o > 0$ may depend on the polynomial Q. Write

$$Q(\xi_1, \xi_2, \ldots, \xi_k, x_{k+1}, \ldots) = Q_1(\xi_1, \ldots, \xi_k) + Q_2(\xi_1, \ldots, \xi_k)$$

where Q_1 is a homogeneous polynomial of order m, and Q_2 is a polynomial of order less than m. The polynomial Q_2 can be rewritten as the sum of finitely many Wiener-Itô integrals with multiplicity less than m. Hence the already proved part of Theorem 6.6 implies that

$$P(|Q_2(\xi_1, \ldots, \xi_k)|>x) \leq \exp(-\bar{K}x^{\frac{2}{m-1}})$$

(we may assume that $m \geq 2$). Then an application of relation (6.9) to Q_1 implies the remaining part of Theorem 6.6, thus it suffices to prove (6.9).

There exist some $\alpha > 0$ and $\beta > 0$ such that

$$\lambda(|Q(\frac{x_1}{|x|}, \ldots, \frac{x_k}{|x|})|>\alpha) > \beta ,$$

where $|x|^2 = \sum_{j=1}^{k} x_j^2$, and λ is the Lebesgue measure on the k-dimensional unit sphere S_{k-1}. Exploiting that $|\xi|$ and $\frac{\xi}{|\xi|}$ are independent, $\frac{\xi}{|\xi|}$ is uniformly distributed on the unit sphere S_{k-1}, and $P(|\xi|>x)\geq C\exp(-x^2)$ for a k dimensional standard normal random variable, we obtain that

$$P(|Q(\xi_1,\dots,\xi_k)|>x)\geq P(|\xi|^m>\frac{x}{\alpha})\ \beta \geq \exp(-Kx^{\frac{2}{m}})$$

if K and x are sufficiently large. Theorem 6.6 is proved.

Theorem 6.6 implies in particular that Wiener-Itô integrals of different multiplicity have a different distribution. A bounded random variable measurable with respect to the σ-algebra generated by a stationary Gaussian field can be expressed as a sum of multiple Wiener-Itô integrals. Another consequence of Theorem 6.6 is the fact that the number of terms in this sum must be infinite. In Theorems 6.2 and 6.2' we have defined a large class of self-similar fields. The question arises whether this class contains self-similar fields such that the distributions of their random variables tend to one (to zero) at infinity (at minus infinity) much faster than the normal distribution function does. This question has been unsolved by now. By Theorem 6.6 such fields, if any, must be expressed as sum of infinitely many Wiener-Itô integrals. The above question is of much greater importance than it may seem at the first instant. Some considerations suggest that in some important models of statistical physics self-similar fields with very fast decreasing tail distributions appear as limit, when the so-called renormalization group transformations are applied for the probability measure describing the state of the model at critical temperature. (The renormalization group transformations are the transformations over the distribution of stationary fields induced by formula (1.1) or (1.3) when $A_N = N^\alpha$, $A(t) = t^\alpha$ with some α.) No rigorous proof about the existence of such self-similar fields is known yet. Thus the real problem behind the above question is whether the self-similar fields interesting for statistical physics can be constructed via multiple Wiener-Itô integrals.

7. On the original Wiener-Itô integral

In this section we briefly describe the definition of the original Wiener-Itô integral as it was done by Itô in [17]. As the arguments are very similar to those of sections 4 and 5 (only the notations become simpler) we omit most proofs.

Let a measure space $(M, \mathcal{M}, \mu)$ with a σ-additive measure μ be given. Let μ satisfy the following continuity property: For all $\varepsilon > 0$ and $A \in \mathcal{M}$, $\mu(A) < \infty$, there exist some disjoint $B_j \in \mathcal{M}$ $m=1,2,\dots,M$ such that $\mu(B_j) < \varepsilon$ and $A = \bigcup_{j=1}^{M} B_j$. The system of random variables $Z_\mu(A)$, $A \in M$, $\mu(A) < \infty$ is called a (Gaussian) random orthogonal measure corresponding to the measure μ if

(i) $Z_\mu(A_1),\dots,Z_\mu(A_k)$ are independent Gaussian random variables if the sets $A_j \in M$, $\mu(A_j) < \infty$, $j = 1,\dots,k$ are disjoint

(ii) $EZ_\mu(A) = 0$, $EZ_\mu(A)^2 = \mu(A)$.

(iii) $Z_\mu(\sum_{j=1}^{k} A_j) = \sum_{j=1}^{k} Z_\mu(A_j)$ with probability 1 if $A_1,\dots,A_k$ are disjoint.

We define the real Hilbert spaces $\bar{K}_\mu^n$, $n=1,2,\dots$. The space consists of the real-valued measurable functions over $\underbrace{M\times\dots\times M}_{n \text{ times}}$, $\underbrace{M\times\dots\times M}_{n \text{ times}}$ such that

$$||f||^2 = \int |f(x_1,\dots,x_n)|^2 \mu(dx_1)\dots\mu(dx_n) < \infty ,$$

and the last formula defines the norm in $\bar{K}_\mu^n$. Let K_μ^n denote the subspace of $\bar{K}_\mu^n$ consisting of the functions $f \in \bar{K}_\mu^n$ such that

$$f(x_1,\dots,x_n) = f(x_{\pi(1)},\dots,x_{\pi(n)}) \quad \text{for all} \quad \pi \in \Pi_n .$$

Let the spaces $\bar{K}_\mu^o$ and K_μ^o consists of the real constants with the norm $||c|| = |c|$. Finally we define the Fock space $\mathrm{Exp}K_\mu$ which consists of the sequences $f = (f_o,f_1,\dots)$ $f_n \in K_\mu^n$, $n=0,1,\dots$, such that

$$||f||^2 = \sum_{n=o}^{\infty} \frac{1}{n!} ||f_n||^2 < \infty .$$

Given a random orthogonal measure Z_μ corresponding to μ on a probability space $(\Omega,\mathcal{A},P)$, let $\mathcal{F} = \sigma(Z_\mu(A),\ A \in \mathcal{M},\ \mu(A) < \infty)$. Let $\mathcal{K}$ denote the real Hilbert space of square integrable random variables measurable with respect to the σ-algebra $\mathcal{F}$. Let $\mathcal{K}_{\le n}$ denote the subspace that is the closure of the linear space containing the polynomials of the random variables $Z_\mu(A)$ of order less than or equal to n. Let $\mathcal{K}_n$ be the orthogonal completition of $\mathcal{K}_{\le(n-1)}$ in $\mathcal{K}_{\le n}$. (The norm is defined as $||\xi||^2 = E\xi^2$ in these Hilbert spaces).

The multiple Wiener-Itô integrals with respect to the random orthogonal measure Z_μ, to be defined below, give a unitary transformation from $\mathrm{Exp}\mathcal{K}_\mu$ to $\mathcal{K}$. We shall denote these integrals by $\int'$ in order to distinguish them from the Wiener-Itô integrals defined in section 4.

First we define the class of elementary functions $\hat{\mathcal{K}}_\mu^n \subset \bar{\mathcal{K}}_\mu^n$. The function $f \in \bar{\mathcal{K}}_\mu^n$ is in $\hat{\mathcal{K}}_\mu^n$ if there exists a finite system of disjoint sets $\Delta_1,\dots,\Delta_N$, $\Delta_j \in \mathcal{M}$, $\mu(\Delta_j) < \infty$, $j=1,2,\dots,N$ such that $f(x_1,\dots,x_n)$ is constant on the set $\Delta_{j_1}\times\dots\times\Delta_{j_n}$ if the indices $j_1,\dots,j_n$ are disjoint and $f(x_1,\dots,x_n)$ equals zero outside these sets. We define

$$\int' f(x_1,\dots,x_n)Z_\mu(dx_1)\dots Z_\mu(dx_n) = \sum f(x_{j_1},\dots,x_{j_n})Z_\mu(\Delta_{j_1})\dots Z_\mu(\Delta_{j_n})$$

for $f \in \hat{\mathcal{K}}_\mu^n$, where $x_k \in \Delta_k$, $k=1,\dots,N$.

Let $\hat{K}^n_\mu = \hat{\bar{K}}^n_\mu \cap K^n_\mu$. The random variables

$$I'_\mu(f) = \frac{1}{n!} \int' f(x_1,\dots,x_n) Z_\mu(dx_1)\dots Z_\mu(dx_n) \ , \quad f \in \hat{\bar{K}}^n_\mu$$

have zero expectation, integrals of different order are orthogonal,

$$I'_\mu(f) = I'_\mu\,(\mathrm{Sym} f), \text{ and } \mathrm{Sym} f \in \hat{K}^n_\mu \text{ if } f \in \hat{\bar{K}}^n_\mu$$

$$(7.1) \qquad E\, I'_\mu(f)^2 \le \frac{1}{n!}\, ||f||^2 \quad \text{if} \quad f \in \hat{\bar{K}}^n_\mu \ ,$$

and (7.1) holds with equality if $f \in \hat{K}^n_\mu$.

It can be seen that $\hat{\bar{K}}^n_\mu$ is dense in $\bar{K}^n_\mu$, hence relation (7.1) enables us to extend the definition of the n-fold Wiener-Itô integrals over $\bar{K}^n_\mu$. All the above mentioned relations remain valid if $f \in \hat{\bar{K}}^n_\mu$ is substituted by $f \in \bar{K}^n_\mu$ and $f \in \hat{K}^n_\mu$ is substituted by $f \in K^n_\mu$. We formulate Itô's formula for these integrals. It can be proved similarly to Theorem 4.2.

Theorem 7.1

Let $\varphi_1,\dots,\varphi_m$, $\varphi_j \in K^1_\mu$, be an orthonormal system in L^2_μ . Let some positive integers $j_1,\dots,j_m$ be given, $j_1+\dots+j_m = N$, and define for all $i=1,\dots,N$

$$g_i = \varphi_s \quad \text{for} \quad j_1+\dots+j_{s-1} < i \le j_1+\dots+j_s$$

Then

$$H_{j_1}(\int' \varphi_1(x) Z_\mu(dx)) \dots H_{j_m}(\int' \varphi_m(x) Z_\mu(dx)) =$$

$$= \int' g_1(x_1) \dots g_N(x_N) Z_\mu(dx_1) \dots Z_\mu(dx_N) =$$

$$= \int' \mathrm{Sym}[g_1(x_1) \dots g_N(x_N)] Z_\mu(dx_1) \dots Z_\mu(dx_N) .$$

(We remark that the diagram formula (Theorem (5.3)) also remains valid for this integral if x_j replaces $-x_j$ in the definition of h_γ.)

It can be seen with the help of Theorem 7.1 that the transformation $I_\mu : \mathrm{Exp}\, K_\mu \to K$, $I_\mu(f) = \sum_{n=0}^{\infty} I_\mu(f_n)$, $f = (f_0, f_1, \dots) \in \mathrm{Exp} K_\mu$ is a unitary transformation, and so are the transformations $(n!)^{1/2} I_\mu$ from K_μ^n to K_n.

Let us consider the special case $(M, M, \mu) = (R^\nu, B^\nu, \lambda)$, where λ denotes the Lebesgue measure in R^ν. A random orthogonal measure corresponding to λ is called a white noise. A random spectral measure corresponding to λ, when the Lebesgue measure is considered as the spectral measure of a generalized field, is also called a white noise. The next result, which can be considered us a random Plancherel formula, establishes a connection between the two types of Wiener-Itô integrals with respect to white noise.

Proposition (7.2)

Let $f = (f_0, f_1, \dots) \in \mathrm{Exp}\, K_\lambda$ a Then $f' = (f'_0, f'_1, \dots) \in \mathrm{Exp} H_\lambda$ with $f'_0 = f_0$ and $f'_n = (2\pi)^{-\frac{n\nu}{2}} \tilde{f}_n$, and

$$\sum\frac{1}{n!}\int{}'f(x_1,\ldots,x_n)Z_\lambda(dx_1)\ldots Z_\lambda(dx_n)\overset{\Delta}{=}\sum\frac{1}{n!}\int f_n'(x_1,\ldots,x_n)Z_\lambda(dx_1)\ldots Z_\lambda(dx_n)$$

Proof of Proposition 7.2

$$(2\pi)^{-\frac{n\nu}{2}}\|\tilde{f}_n\|_{L^2_\lambda} = \|f_n\|_{L^2_\lambda}, \quad \text{hence} \quad f' \in \mathrm{Exp}H_\lambda .$$

Let $\varphi_1, \varphi_2, \ldots$ be a complete orthonormal system in L^2_λ. Then $\varphi_1', \varphi_2', \ldots$ is also a complete orthonormal system in L^2_λ, and if

$$f_n = \sum c_{j_1,\ldots,j_n}\varphi_{j_1}(x_1)\ldots\varphi_{j_n}(x_n)$$

then

$$f_n' = \sum c_{j_1,\ldots,j_n}\varphi'_{j_1}(x_1)\ldots\varphi'_{j_n}(x_n) .$$

Hence Itô's formula for both types of integrals, (i.e. Theorems 4.2 and 7.1) imply Proposition 7.2.

Finally we restrict ourselves to the case $\nu = 1$. We formulate a result which reflects a connection between multiple Wiener-Itô integrals and Itô integrals. Let $W(t)$, $a \le t \le b$, be a Wiener process, and define the random orthogonal measure $Z(dx)$ as

$$Z(A) = \int\chi_A(x)W(dx) , \quad A \subset [a,b), \quad A \in B^1 .$$

Then we have the following

Proposition 7.3

Let $f \in K^n_{\lambda[a,b)}$, where $\lambda[a,b)$ denotes the Lebesgue measure on the interval $[a,b)$. Then

$$\int f(x_1,\dots,x_n)\, Z(dx_1)\dots Z(dx_n) =$$

$$= n!\int_a^b\Big(\int_a^{t_n}\Big(\dots\Big(\int_a^{t_3}\Big(\int_a^{t_2} f(t_1,\dots,t_n)W(dt_1)\Big)W(dt_2)\Big)\dots\; W(dt_n)\Big)$$

Proof of Proposition 7.3

Given an $f \in \hat{K}^n_{\lambda[a,b)}$, let the function $\hat{f}$ be defined as

$$\hat{f}(x_1,\dots,x_n) = \begin{cases} f(x_1,\dots,x_n) & \text{if } x_1<x_2<\dots<x_n \\ 0 & \text{otherwise} \end{cases}$$

It is not difficult to check Proposition 7.3 in the special case $f \in \hat{K}^n_{\lambda[a,b)}$, because of the relation $I(f) = n!I(\hat{f})$. Then a simple limiting procedure proves Proposition 7.3 in the general case.

As a result of Proposition 7.3, in the case $\nu = 1$ multiple Wiener-Itô integrals can be substituted by Itô integrals in the investigation of most problems. In the case $\nu \geq 2$ there is no simple definition of Itô-integrals. On the other hand, no problem arises in generalizing the definition of multiple Wiener-Itô integrals to the case $\nu \geq 2$.

8 Non-central limit theorems

In this section we investigate the problem formulated in section 1, and we show how the technique of Wiener-Itô integrals can be applied for the investigation of such a problem. We restrict ourselves to the case of discrete fields, although the case of generalized fields can be discussed in almost the same way. The proof of some details will be omitted. They can be found in [8]. First we recall the following

Definition 8 A.

The function $L(t)$, $t \in [t_o, \infty)$, $t_o \geq 0$, is said to be slowly varying (at infinity) if

$$\lim_{t\to\infty} \frac{L(st)}{L(t)} = 1 \qquad \text{for all} \quad s > 0 .$$

Theorem 8 A. (Karamata)

If the slowly varying function $L(t)$ is bounded on every finite interval, then it can be represented in the form

$$L(t) = a(t) \exp \left[\int_{t_o}^{t} \frac{\varepsilon(s)}{s} \, ds \right] ,$$

where $a(t) \to a_o \neq 0$ and $\varepsilon(t) \to 0$ as $t \to \infty$.

Let X_n, $n \in Z_\nu$, be a stationary Gaussian field with a correlation function

(8.1) $r(n) = EX_oX_n \sim |n|^{-\alpha} a(\frac{n}{|n|})L(|n|)$,

where $0 < \alpha < \nu$, $L(t)$ is a slowly varying function, bounded in all finite intervals, and $a(t)$ is a continuous function on the unit sphere $S_{\nu-1}$. Let G denote the spectral measure of the field X_n , and let the measures G_N , $N=1,2,\ldots$ be defined by the formula

(8.2) $G_N(A) = \frac{N^\alpha}{L(N)} G(\frac{A}{N})$, $A \in \mathcal{B}^\nu$, $N=1,2,\ldots$

Let G_n , $n=1,2,\ldots$ be a sequence of locally finite measures over R^ν , i.e. $G_n(A) < \infty$ for all measurable bounded sets A . We say that the sequence G_n tends vaguely to a locally finite measure G_o (in notation $G_N \overset{v}{\to} G_o$) if

$$\lim_{n\to\infty} \int f(x)G_n(dx) = \int f(x)G_o(dx)$$

for all continuous functions f with a bounded support. We formulate the following

Lemma 8.1

Let G be the spectral measure of a stationary field with a correlation function $r(n)$ of the form (8.1). Then the sequence of measures G_N defined in (8.2) tends vaguely to a measure G_o . G_o has the homogeneity property

(8.3) $G_o(A) = t^{-\alpha} G_o(tA)$, $A \in \mathcal{B}^\nu$, $t > 0$, and it satisfies the identity

$$2^{\nu}\int e^{i(t,x)}\prod_{j=1}^{\nu}\frac{1-\cos x^{(j)}}{(x^{(j)})^2}G_o(dx) =$$

(8.4)

$$= \int_{[-1,1]^{\nu}} (1-|x^{(1)}|)\dots(1-|x^{(\nu)}|)\,\frac{a\left(\frac{x+t}{|x+t|}\right)}{|x+t|^{\alpha}}\,dx, \quad t\in R^{\nu}\,.$$

We postpone the proof of Lemma 8.1 for a while.

Formulae 8.3 and 8.4 imply that the function $a(t)$ and the number α uniquely determine the measure G_o. Indeed, they determine the measure $\prod_{j=1}^{\nu}\frac{1-\cos x^{(j)}}{(x^{(j)})^2}G_o(dx)$ by (8.4), and hence also the measure G_o by (8.3). Now we formulate

Theorem 8.2

Let X_n, $n\in Z_{\nu}$, be a stationary Gaussian field with a correlation function $r(n)$ defined in (8.1). Let us assume that $0<\alpha<\frac{\nu}{k}$ with some integer k, and define the field $\xi_n = H_k(X_n)$, $n\in Z_{\nu}$. If the fields Z_n^N, $N=1,2,\dots;$ $n\in Z_{\nu}$ are defined by formula (1.1) with $A_N = N^{\nu-\frac{k\alpha}{2}}L(N)^{\frac{k}{2}}$ then their multi-dimensional distributions tend to those of the field Z_n^*,

$$Z_n^* = \int\tilde{\chi}_n(x_1+\dots+x_k)Z_{G_o}(dx_1)\dots Z_{G_o}(dx_k), \quad n\in Z_{\nu}\,.$$

Here Z_{G_o} is a random spectral measure corresponding to the spectral measure G_o which appeared in Lemma 8.1. The function $\tilde{\chi}_n$ is the same as in section 6.

First we explain why the choice of the normalizing constant A_N in Theorem 7.2 was natural, then we explain the idea of the proof, finally we work out the details.

Corollary 5.5 implies, in particular, that $EH_k(\xi)H_k(\eta) = k!(E\xi\eta)^k$ for a Gaussian random vector (ξ,η) with $E\xi = E\eta = 0$, $E\xi^2 = E\eta^2 = 1$. Hence

$$E(Z_n^N)^2 = \frac{1}{A_N^2} \sum_{\substack{j\in B_o^N \\ \ell\in B_o^N}} r(j-\ell)^k \sim \frac{1}{A_N^2} \sum_{j,\ell\in B_o^N} |j-\ell|^{-k\alpha} a\left(\frac{j-\ell}{|j-\ell|}\right)^k (|j-\ell|)^k .$$

Some calculation shows that with our choice of A_N , the expectation $E(Z_n^N)^2$ is separated both from zero and infinity, therefore this is the natural norming factor. In this calculation we have to exploit the condition $k\alpha < \nu$, which implies that in the sum expressing $E(Z_n^N)^2$ those terms are dominant whose arguments $j-\ell$ are large.

The field ξ_n is subordinated to the Gaussian field X_n . It is natural to rewrite it in canonical form, and to express Z_n^N via multiple Wiener-Itô integrals. Itô's formula yields the relation

$$\xi_n = H_k\left(\int e^{i(n,x)} Z_G(dx)\right) = \int \exp[i(n,x_1+\dots+x_k)] Z_G(dx_1)\dots Z_G(dx_k),$$

where Z_G is the random spectral measure adapted to the random field X_n. Then

$$Z_n^N = \frac{1}{A_N} \sum_{j \in B_n^N} \int \exp[i(j, x_1 + \dots + x_k)] \, Z_G(dx_1) \dots Z_G(dx_k) =$$

$$= \frac{1}{A_N} \int \exp[i(Nn, x_1 + \dots + x_k)] \prod_{j=1}^{\nu} \frac{\exp[iN(x_1^{(j)} + \dots + x_k^{(j)})] - 1}{\exp[i(x_1^{(j)} + \dots + x_k^{(j)})] - 1} Z_G(dx_1)$$

$$\dots Z_G(dx_k)$$

Let us make the substitution $y_j = Nx_j$ in the last formula, and let us rewrite it in a form resembling formula (6.5'). To this end, let us introduce the measures G_N defined in (8.2). We can write

$$Z_n^N = \int f_N(y_1, \dots, y_k) \tilde{\chi}(y_1 + \dots + y_k) Z_{G_N}(dy_1) \dots Z_{G_N}(dy_k)$$

with

$$f_N(y_1, \dots, y_k) = \prod_{j=1}^{\nu} \frac{i(y_1^{(j)} + \dots + y_k^{(j)})}{[\exp(i\frac{1}{N}(y_1^{(j)} + \dots + y_k^{(j)})) - 1)]N} .$$

The functions f_N tend to 1 uniformly in all bounded regions, and the measures G_N tend vaguely to G_0 by Lemma 8.1. These relations suggest the following limiting procedure. The limit of Z_n^N can be obtained by substituting f_n with 1 and G_N with G_0 in the Wiener-Itô integral expressing Z_n^N. We want to justify this formal limiting procedure. For this we have to show that the Wiener-Itô integral expressing Z_n^N is essentially concentrated on a large bounded region independent of N. The L_2 isomorphism of Wiener-Itô integrals can help us in showing that. The next lemma is a useful tool for the justification of the above limiting procedure.

Lemma 8.3.

Let G_N, $N=1,2,\dots$ be a sequence of spectral measures on R^ν tending vaguely to a spectral measure G_o. Let a sequence of measurable functions $K_N=K_N(x_1,\dots,x_k)$, $N=0,1,\dots$ be given such that $K_N\in \bar{H}^k_{G_N}$. $N=1,2,\dots$. Assume further the following conditions: For all $\varepsilon>0$ there exist some $A>0$, $N_o>0$ and finitely many rectangulars $P_1,\dots,P_M$, $M=M(\varepsilon)$ on $R^{k\nu}$ such that

a) K_o is continuous on $B=[-A,A]^{k\nu}-\bigcup_{j=1}^{M} P_j$, and $K_N\to K_o$ uniformly on B as $N\to\infty$.

b) $\int_{R^{k\nu}-B} |K_N(x_1,\dots,x_k)|^2\, G_N(dx_1)\dots G_N(dx_k)<\varepsilon$ if $N=0$ or $N>N_o$.

Then $K_o\in \bar{H}^k_{G_o}$, and

$$\int K_N(x_1,\dots,x_k)Z_{G_N}(dx_1)\dots Z_{G_N}(dx_k) \xrightarrow{\mathcal{D}} \int K_o(x_1,\dots,x_k)Z_{G_o}(dx_1)\dots Z_{G_o}(dx_k) \quad \text{as } N\to\infty.$$

Proof of Lemma 8.3.

Conditions a) and b) obviously imply that

$$\int |K_o(x_1,\dots,x_k)|^2 G_o(dx_1)\dots G_o(dx_k)<\infty ,$$

hence $K_o \in H^k_{G_o}$. By using the same argument as in the definition of Wiener-Itô integrals with atomic spectral measure we can reduce the lemma to the case when the spectral measures $G_N, N=0,1,\dots$ are non-atomic.

Let us fix an $\varepsilon>0$, and let $A>0$, $N_o>0$ and the rectangles $P_1,\dots,P_M$ satisfy conditions a) and b) with this ε . Then

$$E[\int [1-\chi_B(x_1,\dots,x_k)]K_N(x_1,\dots,x_k)Z_{G_N}(dx_1)\dots Z_{G_N}(dx_k)]^2 \le$$

(8.5)

$$\le \frac{1}{k!}\int_{R^{k\nu}-B}|K_N(x_1,\dots,x_k)|^2 G_N(dx_1)\dots G_N(dx_k) < \varepsilon$$

for $N=0$ or $N>N_o$, where χ_B is the indicator function of the set B.

Since $B\subset[-A,A]^{k\nu}$, and $G_N \xrightarrow{v} G_o$, hence $G_N \times\dots\times G_N(B) < C(A)$ for all $N=0,1,\dots$. Because of this estimate and the uniform convergence $K_N \to K_o$

$$E[\int (K_N(x_1,\dots,x_k)-K_o(x_1,\dots,x_k))\chi_B(x_1,\dots,x_k)Z_{G_N}(dx_1)\dots$$
$$Z_{G_N}(dx_k)]^2 \le$$

(8.6)

$$\le \frac{1}{k!}\int_B |K_N(x_1,\dots,x_k)-K_o(x_1,\dots,x_k)|^2 G_N(dx_1)\dots G_N(dx_k)<\varepsilon$$

for $N>N_1$ with some $N_1=N_1(A)$.

Because of relations (8.5) and (8.6) in order to prove Lemma 8.3 it is enough to show that

$$\int K_o(x_1,\dots,x_k)\chi_B(x_1,\dots,x_k)Z_{G_N}(dx_1)\dots Z_{G_N}(dx_k) \xrightarrow{\mathcal{D}}$$

$$\int K_o(x_1,\dots,x_k)\chi_B(x_1,\dots,x_k)Z_{G_o}(dx_1)\dots Z_{G_o}(dx_k) .$$

Since $G_N \xrightarrow{v} G_o$, and the function $K_o(x_1,\dots,x_k)\chi(x_1,\dots,x_k)$ can be well approximated by functions from $\hat{\bar{H}}^k_{G_o}$ because of the continuity of K_o over the set B, Lemma 8.3 holds true.

Now we turn to the proof of Theorem 8.2.

<u>The proof of Theorem 8.2</u>.

We want to show that for all integers p , real numbers $c_1,\dots,c_p$ and $n_\ell \in Z_\nu$, $\ell=1,\dots,p$,

$$\sum_{\ell=1}^{p} c_\ell Z^N_{n_\ell} \xrightarrow{\mathcal{D}} \sum_{\ell=1}^{p} c_\ell Z^*_{n_\ell} ,$$

since this relation also implies the convergence of the multi-dimensional distributions. Applying the same calculation as before, we get that

$$\sum_{\ell=1}^{p} c_\ell Z^N_{n_\ell} = \frac{1}{A_N} \sum_{\ell=1}^{p} c_\ell \int \sum_{j\in B^N_{n_\ell}} \exp[i(j,x_1+\dots+x_k)]Z_G(dx_1)\dots Z_G(dx_k),$$

and

$$\sum_{\ell=1}^{p} c_p Z_{n_\ell}^N \triangleq \int K_N(x_1,\dots,x_k) Z_{G_N}(dx_1)\dots Z_{G_N}(dx_k)$$

with

$$K_N(x_1,\dots,x_k)=\frac{1}{N^\nu}\sum_{\ell=1}^{p} c_\ell \sum_{j\in B_{n_\ell}^N} \exp[i(\tfrac{j}{N},x_1+\dots+x_k)] =$$

(8.7)

$$= f_N(x_1,\dots,x_k) \sum_{\ell=1}^{p} c_\ell \tilde{\chi}_{n_\ell}(x_1+\dots+x_k) .$$

Let us define the function

$$K_o(x_1,\dots,x_k) = \sum_{\ell=1}^{p} c_\ell \tilde{\chi}_{n_\ell}(x_1+\dots+x_k)$$

and the measures μ_N on $R^{k\nu}$

(8.8) $$\mu_N(A) = \int_A |K_N(x_1,\dots,x_k)|^2 G_N(dx_1)\dots G_N(dx_k),$$

$$A\in\mathcal{B}^{k\nu}, \quad N=0,1,\dots,$$

where G_o is the vague limit of the measures G_N. To prove Theorem 8.2 it is enough to show that Lemma 8.3 can be applied with these G_N and K_N. (We choose no exceptional rectangles P_j in this application of Lemma 8.3.) Since $G_N \xrightarrow{v} G_o$ and $K_N \to K_o$ uniformly in all bounded regions in $R^{k\nu}$, it is enough to show, beside the proof of Lemma 8.1, that the measures μ_N, $N=1,2,\dots$

tend weakly to the (necessarily finite) measure μ_o (in notation $\mu_N \overset{W}{\rightarrow} \mu_o$), i.e. $\int f(x)\mu_N(dx) \rightarrow \int f(x)\mu_o(dx)$ for all continuous and bounded functions f on $R^{k\nu}$. Then this convergence implies condition b) in Lemma 8.3. Moreover, it is enough to show that there exists some measure $\bar{\mu}_o$ such that $\mu_N \overset{W}{\rightarrow} \bar{\mu}_o$, since then $\bar{\mu}_o$ must coincide with μ_o because of the relations $G_N \overset{V}{\rightarrow} G_o$ and $K_N \rightarrow K_o$ uniformly in all bounded regions. There is a well-known theorem in probability theory about the equivalence between weak convergence of finite measures and the convergence of their Fourier transforms. It would be natural to apply this theorem for proving $\mu_N \overset{W}{\rightarrow} \bar{\mu}_o$. On the other hand, we have the additional information that the measures μ_N, $N=1,2,\dots$ are concentrated on the cubes $[-N\Pi, N\Pi)^{k\nu}$ since G is concentrated on $[-\Pi,\Pi)^{\nu}$. Hence it is more fruitful to apply a version of this theorem, where we can exploit our extra information. We formulate it in the following

Lemma 8.4

Let $\mu_1, \mu_2, \dots$ be a sequence of finite measures on R^{ℓ} such that $\mu_N(R^{\ell} - [-C_N\Pi, C_N\Pi)^{\ell}) = 0$ with some sequence $C_N \rightarrow \infty$. Define the modified Fourier transform

$$\varphi_N(t) = \int_{R^{\ell}} \exp[i(\frac{j}{C_N}, x)]\, \mu_N(dx), \quad t \in R^{\ell}$$

where $j=j(t,N)\in Z_\ell$, and $j=[tC_N]$. (For an $x\in R^\ell$ its integer part $[x]$ is the $n\in Z_\ell$ for which $x^{(j)}-1<n^{(j)}\le\le x^{(j)}$, $j=1,2,\dots,\ell$). If for all $t\in R^\ell$ the sequence $\varphi_N(t)$ tends to a function $\varphi(t)$ continuous at the origin, then μ_N weakly tends to a finite measure μ_o. $\varphi(t)$ is the Fourier transform of μ_o.

We make some comments on the conditions of Lemma 8.4. Let us observe that if the measures μ_N or a part of them are shifted with a vector $2\Pi C_N u$, $u\in Z_\ell$, then their modified Fourier transforms $\varphi_N(t)$ do not change because of the periodicity of the trigonometrical functions $\exp[(i\frac{j}{C_N},x)]$, $j\in Z_\ell$. On the other hand these new measures, which are not concentrated on $[-C_N\Pi,C_N\Pi]^\ell$, have no limit. Lemma 8.4 states that if the measures μ_N are concentrated in the cubes $[-C_N\Pi,C_N\Pi]^\ell$ then the convergence of their modified Fourier transforms defined in Lemma 8.4, which is a weaker condition then the convergence of their Fourier transforms, also implies their convergence to a limit measure.

Proof of Lemma 8.4

The proof is a natural modification of the proof about the equivalence of weak convergence of measures and the convergence of their Fourier transforms. First we show that for all $\varepsilon>0$ there exists some $K=K(\varepsilon)$ such that

$$(8.9) \qquad \mu_N(x \mid x \in R^{\ell}\ ,\ |x^{(1)}|>K)<\varepsilon \quad \text{for all} \quad N \geq 1.$$

As $\varphi(t)$ is continuous at the origin there is some $\delta>0$ such that

$$(8.10) \qquad |\varphi(0,\ldots,0)-\varphi(t,\ldots,0)|<\frac{\varepsilon}{2} \quad \text{if} \quad |t|<\delta\ .$$

We have

$$(8.11) \qquad 0 \leq \mathrm{Re}[\varphi_N(0,\ldots,0)-\varphi_N(t,\ldots,0)] \leq 2\varphi_N(0,\ldots,0)\ .$$

The sequence in the middle term of (8.11) tends to $\mathrm{Re}[\varphi(0,\ldots,0)-\varphi(t,\ldots,0)]$. The right-hand side of (8.11) is bounded since it is convergent. Hence the dominated convergence theorem can be applied, and relation (8.10) implies that

$$\lim_{N\to\infty} \frac{1}{\delta} \int_0^{\delta} \mathrm{Re}[\varphi_N(0,\ldots,0)-\varphi_N(t,\ldots,0)]dt$$

$$= \int_0^{\delta} \frac{1}{\delta} \mathrm{Re}[\varphi(0,\ldots,0)-\varphi(t,\ldots,0)]dt<\varepsilon\ .$$

Hence

$$\frac{\varepsilon}{2} > \lim_{N\to\infty} \mathrm{Re} \int_0^{\delta} \frac{1}{\delta} [\varphi_N(0,\ldots,0)-\varphi_N(t,\ldots,0)]dt =$$

$$= \lim_{N\to\infty} \mathrm{Re} \int \sum_{j=0}^{[\delta C_N]} \frac{1}{\delta C_N} [1-\exp(i\frac{jx^{(1)}}{C_N})]\mu_N(dx) \geq$$

$$\geq \limsup_{N\to\infty} \operatorname{Re} \int_{\{|x^{(1)}|>K\}} \frac{1}{\delta C_N} \sum_{j=0}^{[\delta C_N]} [1-\exp(i\frac{jx^{(1)}}{C_N})]\mu_N(dx) =$$

$$= \limsup_{N\to\infty} \operatorname{Re} \int_{\{|x^{(1)}|>K\}} \{1 - \frac{1}{\delta C_N} \frac{1-\exp[i([\delta C_N]+1)\frac{x^{(1)}}{C_N}]}{1-\exp(i\frac{x^{(1)}}{C_N})}\} \mu_N(dx)$$

with arbitrary $K>0$.

Since the measure μ_N is concentrated in $\{x \mid |x^{(1)}| < C_N\Pi\}$, and

$$\operatorname{Re}\frac{1-\exp[i([\delta C_N]+1)\frac{x^{(1)}}{C_N}]}{1-\exp(i\frac{x^{(1)}}{C_N})} > \frac{2}{\sin\frac{x^{(1)}}{C_N}} > \frac{4}{\Pi}\frac{C_N}{x^{(1)}}$$

if $|x^{(1)}| \leq C_N\Pi$, hence

$$\frac{\varepsilon}{2} \geq \limsup_{N\to\infty} \int_{\{|x^{(1)}|>K\}} [1-\frac{4}{\Pi\delta x^{(1)}}]\mu_N(dx) \geq 2\mu_N(|x^{(1)}|>K)$$

with the choice $K=\frac{8}{\Pi\delta}$. Formula (8.9) is proved.

Applying the same argument to the other coordinates we find that for all $\varepsilon>0$ there exists some $C(\varepsilon)$ such that

$$\mu_N(R^{\ell}-[-C(\varepsilon),\ C(\varepsilon)]^{\ell})<\varepsilon \quad \text{for all} \quad N=1,2,\ldots \ .$$

Consider the usual Fourier transforms

$$\tilde{\varphi}_N(t) = \int_{R^\ell} \exp[i(t,x)]\mu_N(dx) , \quad t\in R^\ell .$$

Then

$$|\varphi_N(t)-\tilde{\varphi}_N(t)|\le 2\varepsilon+ \int_{[-C(\varepsilon),C(\varepsilon)]^\ell} |\exp[i(t,x)]-$$

$$-\exp[i\frac{[tN]}{N},x)]|\mu_N(dx) \le 2\varepsilon+\mu_N(R^\ell)\frac{\ell C(\varepsilon)}{N}$$

for all $\varepsilon>0$. Hence $\tilde{\varphi}_N(t)-\varphi_N(t)\to 0$ as $N\to\infty$, and $\tilde{\varphi}_N(t)\to\varphi(t)$. Then Lemma 8.4 follows from standard theorems on Fourier transforms.

In Theorem 8.2 we apply Lemma 8.4 with $C_N=N$ and $\ell=k\nu$ for the sequence of measures μ_N defined in (8.8). Because of the middle term in (8.7) we can write

$$\varphi_N(t_1,\dots,t_k) = \sum_{r=1}^{p} \sum_{s=1}^{p} c_r c_s \psi_N(t_1+n_r-n_s,\dots,t_k+n_r-n_s)$$

with

$$\psi_N(t_1,\dots t_k)=\frac{1}{N^{2\nu}} \int\exp[i\frac{1}{N}((j_1,x_1)+\dots+(j_k,x_k))]$$

$$\sum_{p\in B_0^N} \sum_{q\in B_0^N} \exp[i(\frac{p-q}{N}, x_1+\dots+x_k)]G_N(dx_1)\dots G_N(dx_k)=$$

$$=\frac{1}{N^{2\nu-k\alpha}L(N)^k} \sum_{p\in B_0^N} \sum_{q\in B_0^N} r(p-q+j_1)\dots r(p-q+j_k)$$

where $j_p=[t_pN]$, $t_p \in R^\nu$, $p=1,\dots,k$.
The asymptotical behaviour of $\psi_N(t_1,\dots,t_k)$ for $N\to\infty$ can be investigated by the help of the last relation and formula (8.1). Rewriting the last double sum in the form of a singe sum, one gets

$$\psi_N(t_1,\dots,t_k) = \int_{[-1,1]^\nu} f_N(t_1,\dots,t_k,x)dx$$

with

$$f_N(t_1,\dots,t_k,x)=(1-\frac{[x^{(1)}N]}{N})\dots(1-\frac{[x^{(\nu)}N]}{N})\frac{r([xN]+j_1)}{N^{-\alpha}L(N)}\dots$$

$$\dots\frac{r([xN]+j_k)}{N^{-\alpha}L(N)}$$

with $j_p=[t_pN]$, $p=1,\dots,k$.
It can be seen with the help of formula (8.1) that

$$f_N(t_1,\dots,t_k,x) \to f_o(t_1,\dots,t_k,x)$$

with

$$f_o(t_1,\dots,t_k,x)=(1-|x^{(1)}|)\dots(1-|x^{(\nu)}|)\frac{a(\frac{x+t_1}{|x+t_1|})}{|x+t_1|^\alpha}\dots$$

$$\dots\frac{a(\frac{x+t_k}{|x+t_k|})}{|x+t_k|^\alpha}$$

uniformly on the set $x \in [-1,1]^{\nu} - \bigcup_{p=1}^{k} \{x: |x+t_p| > \varepsilon\}$ for all $\varepsilon > 0$.

We claim that

$$\psi_N(t_1,\dots,t_k) \to \psi_o(t_1,\dots,t_k) = \int_{[-1,1]^{\nu}} f_o(t_1,\dots,t_k,x)\,dx ,$$

and ψ_o is a continuous function.

This relation implies that $\mu_N \xrightarrow{w} \mu_o$. To prove it, it is enough to show that

$$(8.12) \quad \left|\int_{|x+t_k|<\varepsilon} f_o(t_1,\dots,t_k,x)\,dx\right| < C(\varepsilon), \quad p=1,\dots,k$$

and

$$(8.12') \quad \left|\int_{|x+t_p|\le\varepsilon} f_N(t_1,\dots,t_k,x)\,dx\right| < C(\varepsilon) \quad p=1,\dots,k;\ N=1,2,\dots,$$

where $C(\varepsilon) \to 0$ as $\varepsilon \to 0$.

By Hölder's inequality

$$\left|\int_{|x+t_p|\le\varepsilon} f_o(t_1,\dots,t_k,x)\,dx\right| \le C \prod_{\ell=1}^{k} \left[\int_{|x+t_p|\le\varepsilon} |x+t_\ell|^{-\alpha k}\,dx\right]^{1/k} \le$$

$$\le C' \varepsilon^{\nu-\alpha k}$$

with some appropriate $C>0$ and $C'>0$ since $\nu-\alpha k>0$, and $a(.)$ is a bounded function. Similarly

$$\left|\int_{|x+t_p|\le\varepsilon} f_N(t_1,\dots,t_k,x)\,dx\right| \le \prod_{\ell=1}^{k} \left[\int_{|x+t_p|\le\varepsilon} \frac{|r([xN]+j_\ell)|^k}{N^{-k\alpha} L(N)^k}\,dx\right]^{1/k} .$$

It is not difficult to see, by using Karamata's theorem, that if L is a slowly varying function which is bounded in all finite intervals then for all $T>0$ and $\eta>0$ there exists some $C=C(T,\eta)$ such that

$$L(tN) \leq Ct^{-\eta}L(N) \quad \text{for all} \quad t<T \quad \text{and} \quad N\geq 1 .$$

Hence formula (8.1) implies that

$$(8.13) \qquad |r([xN]+j_\ell)| \leq CN^{-\alpha}L(N)|x+t_\ell|^{-\alpha-\eta}$$

and

$$\int_{|x+t_p|<\varepsilon} \frac{|r([xN]+j_\ell)|^k}{N^{-k\alpha}L(N)^k}\, dx \leq B\int_{|x+t_p|<\varepsilon} |x+t_\ell|^{-k(\alpha+\eta)}dx$$

with some $B=B(t_\ell,\eta)$. (Let us remark that (8.13) holds also for $|[xN]+j_\ell| \leq K_1$ with some $K_1>0$ independent of N, because $|r(n)| \leq 1$ for all $n\in Z_\nu$). Therefore we get, by chosing an $\eta>0$ chosing that $k(\alpha+\eta)<\nu$, the inequality

$$|\int_{|x+t_p|<\varepsilon} f_N(t_1,\dots,t_k,x)dx| < C' \prod_{\ell=1}^{k} [\int_{|x+t_p|\leq\varepsilon} |x+t_\ell|^{-k(\alpha+\eta)} dx]^{1/k} \leq$$

$$\leq C''\varepsilon^{\nu-k(\alpha+\eta)} .$$

The right-hand side of this inequality tends to zero as $\varepsilon\to 0$. Hence we proved formulae (8.13) and (8.13') therefore also the relation $\mu_N \overset{w}{\to} \mu_0$. To complete the proof of Theorem 8.2 it remains to prove Lemma 8.1.

<u>Proof of Lemma 8.1</u>

Introduce the notations

$$K_N(x) = \prod_{j=1}^{\nu} \frac{\exp(ix^{(j)})-1}{N[\exp(i\frac{x^{(j)}}{N})-1]}, \qquad N=1,2,\ldots$$

and

$$K_o(x) = \prod_{j=1}^{\nu} \frac{\exp(ix^{(j)})-1}{ix^{(j)}} .$$

Let us consider the measures μ_N defined in formula (8.8) in the special case $k=1,\ p=1,\ c_1=1$. Then

$$\mu_N(A) = \int_A |K_N(x)|^2 G_N(dx) .$$

We have already seen in the proof of Theorem 8.2 that $\mu_N \xrightarrow{w} \mu_o$ with some finite measure μ_o, and the Fourier transform of μ_o is

$$\varphi(t) = \int_{[-1,1]^\nu} (1-|x^{(1)}|)\ldots(1-|x^{(\nu)}|) \frac{a(\frac{x+t}{|x+t|})}{|x+t|^\alpha} dx .$$

First we show that

(8.14) $\qquad \lim_{N\to\infty} \int f(x)G_N(dx) \quad$ exists

for all continuous functions f with compact support.

Let the continuous function f vanish outside the

cube $[-T\Pi, T\Pi]^\nu$, $T\geq 1$. Let $M=[\frac{N}{2T}]$. Then

$$\int f(x)G_N(dx)=\frac{N^\alpha}{L(N)}\ \frac{L(M)}{M^\alpha}\ \int f(\frac{N}{M}x)G_M(dx) =$$

$$=\frac{N^\alpha L(M)}{M^\alpha L(N)}\ \int f(\frac{N}{M}x)|K_M(x)|^{-2}\ \mu_M(dx)\to(2T)^\alpha\int f(2Tx)|K_o(x)|^{-2}\mu_o(dx)$$

because $f(\frac{N}{M}x)|K_M(x)|^{-2}$ vanishes outside the cube $[-\Pi,\Pi]^\nu$, $f(\frac{N}{M}x)|K_M(x)|^{-2}\to f(2Tx)|K_o(x)|^{-2}$ uniformly, and $\mu_M \overset{w}{\to} \mu_o$ as $N\to\infty$. Hence (8.14) is proved. It is not difficult to see that relation (8.14) implies the existence of a locally finite measure G_N such that $G_N \overset{v}{\to} G_o$. As $G_N \overset{v}{\to} G_o$ and $|K_N(x)|^2\to|K_o(x)|^2$ uniformly in all bounded regions, hence $\mu_N \overset{v}{\to} \bar{\mu}_o$, where $\bar{\mu}_o(A) = \int_A|K_o(x)|^2G_o(dx)$, $A\in\mathcal{B}^\nu$. Since $\mu_N \overset{w}{\to} \mu_o$ the measures μ_o and $\bar{\mu}_o$ must coincide, i.e.

$$\mu_o(A) = \int_A|K_o(x)|^2G_o(dx), \qquad A\in\mathcal{B}^\nu .$$

Relation (8.4) expresses the fact that φ_o is the Fourier transform of μ_o.
Let us extend the definition of the measures G_N in (8.2) to all non-negative real numbers N. It is easy to see that the relation $G_t\to G_o$ as $t\to\infty$ remains valid.
Hence

$$\int f(x)G_o(dx) = \lim_{u\to\infty} \int f(x)G_u(dx) =$$

$$= \lim_{u\to\infty} \frac{s^{\alpha} L(\frac{u}{s})}{L(u)} \int f(sx) G_{\frac{u}{s}}(dx) = s^{\alpha} \int f(sx) G_o(dx)$$

for all $s>0$ and continuous functions f with compact support. This identity implies the homogeneity property (8.3) of G_o. Lemma 8.3 is proved.

The next result is a generalization of Theorem 8.2.

Theorem 8.2'

Let X_n, $n \in Z_\nu$, be a stationary Gaussian field with a correlation function $r(n)$ defined in (8.1). Let $H(x)$ be a real function with the properties $EH(X_n)=0$, $EH(X_n)^2 < \infty$. Let us consider the Fourier expansion

$$H(x) = \sum_{j=1}^{\infty} c_j H_j(x) \qquad \sum c_j^2 \, j! < \infty$$

of the function H by the Hermite polynomials H_j. Let k be the smallest index in this expansion such that $c_k \neq 0$. If $0<k\alpha<\nu$ in (8.1), and the field Z_n^N is defined by the field $\xi_n = H(X_n)$, $n \in Z_\nu$ and formula (1.1) then the multi-dimensional distributions of the fields Z_n^N with $A_N = N^{\nu - \frac{k\alpha}{2}} L(N)^{\frac{k}{2}}$ tend to those of the field $c_k Z_n^*$, $n \in Z_\nu$, where the field Z_n^* is the same as in Theorem 8.2.

Proof of Theorem 8.2'

Define $H'(x) = \sum_{j=k+1}^{\infty} c_j H_j(x)$ and $Y_n^N = \frac{1}{A_N} \sum_{\ell \in B_n^N} H'(X_\ell)$.

Because of Theorem 8.2, in order to prove Theorem 8.2' it is enough to show that

$$E(Y_n^N)^2 \to 0 \quad \text{as} \quad N\to\infty .$$

It follows from Corollary 5.5 that

$$E(Y_n^N)^2 = \frac{1}{A_N^2} \sum_{j=k+1}^{\infty} c_j^2 \, j! \sum_{s,t\in B_n^N} [r(t-s)]^j .$$

Hence a simple calculation with the help of formula (8.1) yields

$$E(Y_n^N)^2 = \frac{1}{A_N^2} [O(N^{2\nu-(k+1)\alpha} L(N)^{k+1}) + O(N^{\nu})] \to 0 .$$

Theorem 8.2' is proved.

In Theorems 8.2 and 8.2' we investigated some very special subordinated fields. The next result shows that the same limiting field as the one in Theorem 8.2, appears in a much more general situation.

Let us define the field

$$(8.15) \quad \xi_n = \sum_{j=k}^{\infty} \frac{1}{j!} \int \exp[i(n,x_1+\dots+x_j)]\alpha_j(x_1,\dots,x_j) Z_G(dx_1)\dots Z_G(dx_j), \quad n\in Z_\nu ,$$

where Z_G is the random spectral measure adapted to a Gaussian field X_n, $n \in Z_\nu$, with a correlation function satisfying (8.1) with $0<\alpha<\frac{\nu}{k}$.

Theorem 8.5

Let the fields Z_n^N be defined by formulae (8.15) and (1.1) with $A_N = N^{\nu - \frac{k\alpha}{2}} L(N)^{\frac{k}{2}}$. The multi-dimensional distributions of the fields Z_n^N tend to those of the field $\alpha_k(0,\dots,0) Z_n^*$, where the field Z_n^* is the same as in Theorem 8.2, if only the following conditions are fulfilled

(i) $\alpha_k(x_1,\dots,x_k)$ is a bounded function, continuous at the origin, and such that $\alpha_k(0,\dots,0) \neq 0$.

(ii) $$\sum_{j=k+1}^{\infty} \frac{1}{j!} \frac{N^{-(j-k)\alpha}}{L(N)^{j-k}} \int_{R^{j\nu}} |\alpha_j(\frac{x_1}{N},\dots,\frac{x_j}{N})|^2 \frac{1}{N^{2\nu}}$$

$$|\sum_{\ell \in B_0^N} \exp[i(\frac{\ell}{N}, x_1+\dots+x_j)]|^2 G_N(dx_1)\dots G_N(dx_j) \to 0,$$

where G_N is defined in (8.2).

Proof of Theorem 8.5

The proof is very similar to those of Theorems 8.2 and 8.2'. The same argument as in the proof of Theorem 8.2' shows that because of condition (ii) ξ_n can be substituted in the present proof by the following

$$\xi_n' = \frac{1}{k!}\int \exp[i(n, x_1+\dots+x_k)]\alpha_k(x_1,\dots,x_k) Z_G(dx_1)\dots Z_G(dx_k),$$
$$n\in Z_\nu ,$$

Then a natural modification in the proof of Theorem 8.2 implies Theorem 8.5. The main point in this modification is that we have to substitute the measures μ_N defined in formula (8.8) by the measures $\bar{\mu}_N$:

$$\bar{\mu}_N(A) = \int_A |K_N(x_1,\dots,x_k)|^2 |\alpha_k(\frac{x_1}{N},\dots,\frac{x_k}{N})|^2 \; G_N(dx_1)\dots G_N(dx_k),$$
$$A\in B^{k\nu}$$

and to observe that, because of condition (i) the limit relation $\mu_N \overset{w}{\to} \mu_0$ implies that $\bar{\mu}_N \overset{w}{\to} |\alpha_k(0,\dots,0)|^2 \mu_0$.

The content of Theorem 8.5 is the following: Let ξ_n, $n\in Z_\nu$, be a field subordinated to a Gaussian field X_n, $n\in Z_\nu$, with a correlation function satisfying formula (8.1). Let us write the field ξ_n in its canonical form (6.1). As $N\to\infty$, the limit behaviour of the fields Z_n^N defined by formula (1.1) and the field ξ_n , depends on the smallest index with a non-vanishing term in the canonical representation of ξ_n. If this index is k , and $0<k\alpha<\nu$, then the limit of the fields Z_n^N with the normalizing constants $A_N = N^{\nu-\frac{k\alpha}{2}} L(N)^{\frac{k}{2}}$ is the self-similar field Z_n^* defined in Theorem 8.2 by k-fold

Wiener-Itô integrals. If $k\alpha \geq \nu$ then the dependence between distant terms of the field ξ_n is relatively small. Under some slight conditions the fields Z_n^N coverge in this case to a field of independent normal random variables. The normalizing factor must be chosen as $A_N = N^{\frac{\nu}{2}}$ in general, but in the case $k\alpha = \nu$ there are examples when the good normalizing constant is $A_N = N^{\frac{\nu}{2}} \bar{L}(N)$ with a non-constant slowly varying function $\bar{L}$. These results can be proved, with the help of Corollary 5.5, by showing that the moments of a linear combination $\sum c_p Z_{n_p}$ tend to those of a normal random variable. We omit the proof.

The main problem in applying Theorem 8.5 is to check conditions (i) and (ii). We remark without proof that any field of form $\xi_n = H((X_{s_1+n}, \dots, X_{s_{p+n}})$ $s_1, \dots, s_p$, $n \in Z_\nu$, satisfies condition (ii). This is proved in Remark 6.2 of [8]. If the conditions (i) or (ii) are violated, then a limit of different type may appear. Finally, we quote such a result without proof. (See [21] for a proof.) Here we restrict ourselves to the case $\nu = 1$. The limiting field appearing in this result belongs to the class of self-similar field constructed in Remark 6.5.

Let a_n, $n = \dots -1, 0, 1 \dots$ be a sequence of real numbers such that

$$(8.16) \quad \begin{array}{ll} a_n = C(1) n^{-\beta-1} + o(n^{-\beta-1}) & \text{if } n \geq 0 \\ a_n = C(2) |n|^{-\beta-1} + o(|n|^{-\beta-1}) & \text{if } n < 0 \end{array} \qquad -1 < \beta < 1 .$$

Let X_n, $n=\ldots-1,0,1,\ldots$ be a stationary Gaussian sequence with correlation function $r(n)=EX_0X_n=|n|^{-\alpha}L(|n|)$ $0<\alpha<1$, where $L(.)$ is a slowly varying function. Define the field ξ_n, $n=\ldots-1,0,1,\ldots$ as

$$\xi_n = \sum_{m=-\infty}^{\infty} a_m H_k(Y_{m+n}) . \tag{8.17}$$

Theorem 8.6

Let the sequence ξ_n, $n=\ldots-1,0,1\ldots$ be defined by (8.16) and (8.17). Let $0<k\alpha<1$, $0<1-\beta-\frac{k}{2}\alpha<1$, and let one of the following conditions be satisfied

(a) $0<\beta<1$ and $\sum_{n=-\infty}^{\infty} a_n = 0$

(b) $0>\beta>-1$

(c) $\beta=0$, $C(1)=-C(2)$, and $\sum_{n=0}^{\infty} |a_n+a_{-n}|<\infty$.

Let us define the sequences Z_n^N by formula (1.1) with the choice $A_N=N^{1-\beta-\frac{k}{2}\alpha}L(N)^{\frac{k}{2}}$ and the above defined field ξ_n. The multi-dimensional distributions of the sequences Z_n^N tend to those of the sequence $D^{-k}Z_n^*=D^{-k}Z_n^*(\alpha,\beta,k,b,c)$, where

$Z_n^*(\alpha,\beta,k,b,c) =$

$$=\int \tilde{\chi}_n(x_1+\dots+x_k)[b|x_1+\dots+x_k|^{\beta}+ci|x_1+\dots+x_k|^{\beta}\operatorname{sign}(x_1+\dots+x_k)]$$

$$|x_1|^{\frac{\alpha-1}{2}}\dots|x_k|^{\frac{\alpha-1}{2}}\,W(dx_1)\dots W(dx_k)\;,$$

$W(.)$ denotes the white noise field, i.e. a random spectral measure corresponding to the Lebesgue measure,

$D=2\Gamma(\alpha)\cos(\frac{\alpha}{2}\pi)$, $b=2[C(1)+C(2)]\Gamma(-\beta)\sin(\frac{\beta+1}{2}\pi)$,

$c=2[C(1)-C(2)]\Gamma(-\beta)\cos(\frac{\beta+1}{2}\pi)$ in cases (a) and (b),

and $b=\sum_{n=-\infty}^{\infty} a_n$, $c=C(1)$ in case (c).

9. History of the problems. Comments.

Section 1.

In statiscical physics the problem formulated in this section appeared at the investigation of some physical models at critical temperature. A discussion of this problem and further references can be found in the fourth chapter of the forthcoming book of Ja.G.Sinai [31]. The first example of a limit theorem for partial sums of random variables which is considerably different from the independent case was given by M.Rosenblatt in [26]. Further results in this direction were proved by R.L.Dobrushin, H.Kesten and F.Spitzer, P.Major, M.Rosenblatt and M.S.Taqqu [6], [7], [8], [18], [21], [27], [28], [32], [35]. In most of these paper only the one-dimensional case is considered, and it is formulated in a different but equivalent way. The joint distribution of the random variables $A_N^{-1} \sum_{j=1}^{[Nt]} \xi_j$, $0<t<\infty$, is investigated.

Similar problems also appeared in the theory of infinite particle systems. The large-scale limit of the so-called voter model and of infinite particle branching Brownian motions were investigated in papers [2],[5], [16], [22]. It was proved that in these models the limit is, with a non-typical normalization, a Gaussian self-similar field. The investigation of the large-scale limit would be very natural for many other infinite particle systems, but in most cases this problem is hopelessly difficult.

The notion of subordinated fields in the present context first appeared at Dobrushin [6]. It is natural to expect that there exists a large class of self-similar fields which cannot be obtained as subordinated fields. Nevertheless the present techniques are not powerful enough for finding them.

The approach to the problem is different in statistical physics. In statistical physics one looks for self-similar fields which satisfy some conditions formulated in accordance to physical considerations. One tries to describe these fields with the help of a power series which is to define the Radon-Nykodym derivative of the field with respect to a Gaussian field. The deepest result in this direction is a recent paper of P.M.Bleher and M.D.Missarov [1] who can define the required formal power series. This result enables one to calculate several critical indices interesting for physisists, but the task of proving that this formal expression defines an existing field seems to be very hard. It is also an open problem whether the class of self--similar fields constructed via multiple Wiener-Itô integrals contains the non-Gaussian self-similar fields interesting for statistical physics. Some experts are very sceptical in this respect. The Gaussian self-similar fields are investigated in [6] and [30]. A more thorough investigation is under preparation in [8].

The notion of generalized fields was introduced by I.M.Gelfand. A detailed discussion can be found in the book [14], where the properties of Schwartz spaces we need can also be found.

In the definition of generalized fields the class of test functions $\mathcal{S}$ can be substituted by other linear topological spaces consisting of real valued function. The most frequently considered space, beside the space $\mathcal{S}$, is the space $\mathcal{D}$ of infinitely many times differentiable functions with compact support. In paper [6] Dobrushin also considered the space $\mathcal{S}^r\subset\mathcal{S}$ which consists of the functions $\varphi\in\mathcal{S}$ satisfying the relation $\int x^{(1)j_1}\dots x^{(\nu)j_\nu}\varphi(x)dx=0$ provided that $j_1+\dots+j_\nu<r$. He considered this class of test functions because there are much more continuous linear

functional over S^r than over S, and this property of S^r can be exploited at certain investigations. Generally, no problems arise in the proofs if the space of test functions S is substituted by S^r or $\mathcal{D}$ in the definition of generalized fields.

Two generalized fields $X(\varphi)$ and $\overline{X}(\varphi)$ can be identified if $X(\varphi)\stackrel{\Delta}{=}\overline{X}(\varphi)$ for all $\varphi\in S$. We remark that this relation also implies that the multi-dimensional distributions of the random vectors $(X(\varphi_1),\dots,X(\varphi_n))$ and $(\overline{X}(\varphi_1),\dots,\overline{X}(\varphi_n))$, $\varphi_1,\dots,\varphi_n\in S$, coincide. As S is a linear space, this relation can be deduced from property a) of generalized fields by exploiting that two distribution functions on R^n agree if and only if their characteristic functions agree.

Let S' denote the space of continuous linear functionals over S, and let $\mathcal{A}_S$ be the σ-algebra over S' generated by the sets $A(\varphi,a)=\{F: F\in S',\ F(\varphi)<a\}$, where $\varphi\in S$ and $a\in R^1$ are arbitrary. Given a probability space $(S', \mathcal{A}_{S'}, P)$, the generalized field $\overline{X}=\{\overline{X}(\varphi),\ \varphi\in S\}$ can be defined on it by the formula $\overline{X}(\varphi)(F)=F(\varphi)$, $\varphi\in S$, $F\in S'$. The following deep result is due to Minlos (see e.g.[14]).

<u>Theorem</u> (Minlos)

<u>Let $\{X(\varphi),\ \varphi\in S\}$ be a generalized field. There exists a probability measure P on the measureable space $(S', \mathcal{A}_S)$ such that the generalized field $\overline{X}=\{\overline{X}(\varphi),\ \varphi\in S\}$ defined on the probability space $(S', \mathcal{A}_{S'}, P)$ by the</u>

formula $\overline{X}(\varphi)(F)=F(\varphi)$, $\varphi\in S$, $F\in S'$, satisfies the relation $X(\varphi)\overset{\Delta}{=}\overline{X}(\varphi)$ for all $\varphi\in S$.

The generalized field $\overline{X}$ has some nice properties. Namely, property a) in the definition of generalized fields holds for all $F\in S'$. Moreover, $\overline{X}$ satisfies the following strenghtened version of property b):

b') $\lim\overline{X}(\varphi_n)=\overline{X}(\varphi)$ everywhere on S' if $\varphi_n\to\varphi$ in the topology of S.

Because of this nice behaviour of the field $\overline{X}(\varphi)$ most authors define generalized fields as the versions $\overline{X}$ defined by Minlos' theorem. Since we have never needed the extra properties of the field $\overline{X}$, we have deliberately avoided the application of Minlos' theorem in the definition of generalized field. Minlos' theorem heavily depends on some topological properties of S, namely that S is a so-called nuclear space. Minlos' theorem also holds if the space of test functions is substituted by $\mathcal{D}$ or S^r in the definition of generalized fields.

Section 2.

Wick polynomials are widely used in the literature of statistical physics. A detailed discussion about Wick polynomials can be found in [11]. Theorems 2A and 2B are well-known, and they can be found in the standard literature. Theorem 2C can be found e.g. in Dynkin's book [12] (Lemma 1.5). Theorem 2.1 is due to Segal [29]. It is closely related to a result of Cameron and Martin [3]. The remarks at the end of the section about the content of formula 2.1 are related to [23].

Section 3.

Random spectral measures were independently introduced by Cramer and Kolmogorov [4], [19] . They could have been introduced by means of Stone's theorem about the spectral representation of one-parameter group of unitary operators. Bochner's theorem can be found in any standard text book on functional analysis, the proof of the Bochner-Schwartz theorem can be found in [14]. We remark that the same result holds true if the space of test functions $\mathcal{S}$ is substituted by $\mathcal{D}$.

Section 4.

The stochastic integral defined in this section is a version of that introduced by Itô in [17]. This modified integral first appeared in Totoki's lecture note [36] in a special form. Most results of this section can be found in Dobrushin's paper [6]. The definition of Wiener-Itô integrals in the case when the spectral measure may have atoms in new.

Section 5.

Proposition 5.1 is proved for the original Wiener-Itô integrals by Itô in [17]. Lemma 5.2 contains a well-known formula about Hermite polynomials. The main result of this section, Theorem 5.3, appeared in Dobrushin's work [6].

The proof given there is not complete, several non-trivial details are omitted. Theorem 5.3 is closely related to Feynman's diagram formula. The result of Corollary 5.5 was already known at the beginning of the century. It was proved with the help of some formal manipulations. This formal calculation was justified by Taqqu in [33] with the help of some deep inequalities.

We could not find results similar to Propositions 5.6 and 5.7 in the literature of probability theory. On the other hand, such results are well-known in statistical physics, and they play an important role in constuctive field theory. A sharpened form of these results is Nelson's deep hypercontractive inequality [25], which we formulate below.

Let X_t, $t\in T$, and $Y_{t'}$, $t'\in T'$, be two sets of jointly Gaussian random variables on some probability spaces $(\Omega,\mathcal{A},P)$ and $(\Omega',\mathcal{A}',P')$. Let $\mathcal{H}_1$ and $\mathcal{H}_1'$ be the Hilbert spaces generated by the finite linear combinations $\sum c_j X_{t_j}$ and $\sum c_l Y_{t_j'}$. Let us define the σ-algebras $\mathcal{B}=\sigma(X_t,\ t\in T)$, $\mathcal{B}'=\sigma(Y_{t'},\ t'\in T')$ and the Banach spaces $L_p(X)=L_p(\Omega,\mathcal{B},P)$, $L_p(Y)=L_p(\Omega',\mathcal{B}',P')$, $1\le p\le\infty$. Let A be a linear transformation from $\mathcal{H}_1$ to $\mathcal{H}'_1$ with norm not exceeding 1. We define an operator $\Gamma(A): L_p(X)\to L_{p'}(Y)$ for all $1\le p,\ p'\le\infty$ in the following way: If η is a homogeneous polynomial of the variables X_t

$$\eta = \sum c_{j_1,\dots,j_s}^{t_1,\dots,t_s} X_{t_1}^{j_1} \dots X_{t_s}^{j_s}, \quad t \in T, \quad \text{then}$$

$$\Gamma(A):\eta: = \sum c_{j_1 \dots j_s}^{t_1 \dots t_s} :(AX_{t_1})^{j_1} \dots (AX_{t_s})^{j_s}: \ .$$

It can be proved that this definition is meaningful, i.e. $\Gamma(A):\eta:$ does not depend on the representation of η, and $\Gamma(A)$ can be extended to a bounded operator from $L_1(X)$ to $L_1(Y)$ in a unique way. This means in particular that $\Gamma(A)\xi$ is defined for all $\xi \in L_p(X)$, $p \geq 1$. Nelson's hypercontractive inequality says the following: Let A be a contraction from H_1 to H_1'. Then $\Gamma(A)$ is a contraction from $L_q(X)$ into $L_p(Y)$ for $1 \leq q \leq p$ provided that

$$(+) \ ||A|| \leq \left(\frac{q-1}{p-1}\right)^{\frac{1}{2}} .$$

If $(+)$ does not hold then $\Gamma(A)$ is not a bounded operator from $L_q(X)$ to $L_p(Y)$.

A further generalization of this result can be found in [15].

The following discussion may help to understand the relation between Nelson's hypercontractive inequality and Corollary 5.6. Let us apply Nelson's inequality in the special case when $\{X_t, \ t \in T\} = \{Y_{t'}, \ t' \in T'\}$ is a stationary Gaussian field with spectral measure G; $q=2$, $p=2m$ with some positive integer m, $A=cId$, where Id denotes the identity operator, and $c=(2m-1)^{-\frac{1}{2}}$. Let H^c

H_n^c be the complexifications of the real Hilbert spaces H and H_n defined in section 2. Then $L_2(X)=H^c=H_o^c+H_1^c+ \dots$ by Theorem 2.1 and formula (2.1). The operator $\Gamma(cId)$ equals $c^n Id$ on the subspace H_n^c. If $h_n \in H_G^n$ then $I_G(h_n) \in H_n$, hence Nelson's inequality shows that

$$[EI_G(h_n)^{2m}]^{\frac{1}{2m}} = c^{-n}[E(\Gamma(cId)I_G(h_n))^{2m}]^{\frac{1}{2m}} \le c^{-n}[EI_G(h_n)^2]^{1/2}$$

i.e.

$$EI_G(h_n)^{2m} \le (2m-1)^{mn}(EI_G(h_n)^2)^m .$$

This inequality is very similar to the second inequality in Corollary 5.6, only the multiplying constants are different. Moreover, for large m these multiplying constants are near to each other. We remark that the following weakened form of Nelson's inequality could be deduced relatively easily from Corollary 5.6. Let $A: H_1 \to H_1'$ be a contraction $||A|| = c < 1$. Then there exists a $\bar{p} = \bar{p}(c) > 2$ such that $\Gamma(A)$ is a bounded operator from $L_2(X)$ into $L_p(Y)$ for $p < \bar{p}$. This weakened form of Nelson's hypercontractive inequality is sufficient in many applications.

Section 6.

Theorems 6.1, 6.2 and Corollary 6.4 were proved by Dobrushin in [6]. Taqqu proved similar results [34], but he gave a different representation. Theorem 6.6 was proved by H.P.Mc.Kean in [24]. The proof of the lower bound uses some ideas from [13]. Remark 6.5 is from [21]. As Proposition 6.3 also indicates some non-trivial problems about the convergence of certain integrals must be solved when constructing self-similar fields. Such convergence problems are common in statistical physics. To tackle such problems the so-called power counting method (see e.g. [20]) was worked out. This method could also be applied in this section. Part b) of Proposition 6.3 implies that the self-similarity parameter α cannot be chosen in a larger domain in Corollary 6.4. One can ask about the behaviour of the self-similar fields X_j and $X(\varphi)$ defined in Corollary 6.4 if the self-similarity parameter α tends to the critical value $\frac{\nu}{2}$. The variance of the random variables X_j and $X(\varphi)$ tend to infinity in this case, and the fields X_j, $j\in Z_\nu$, and $X(\varphi)$, $\varphi\in S$, tend, after an appropriate normalization, to a field of independent normal random variables in the discrete and to a white noise field in the continuous case. The proof of these results with a more detailed discussion will appear in [9].

In a recent paper [18] Kesten and Spitzer have proved a limit theorem, where the limit field is a self-similar field which seems not to belong to the class of self-similar fields constructed in section 6. (We cannot, however, exclude the possibility that there exists some self-similar field in the class defined in Theorem 6.2 with the same distribution as this field, although it is given by a completely different form.) This self-similar field constructed by Kesten and Spitzer is the only rigorously constructed self-similar field known for us that does not belong to the fields constructed in Theorem 6.2. We describe this field and then we make some comments.

Let $B_1(t)$ and $B_2(t)$, $-\infty<t<\infty$, be two independent Wiener processes. (We say that $B(t)$ is a Wiener process on the real line if $B(t)$, $t\geq 0$ and $B(-t)$, $t\geq 0$ are two independent Wiener processes.) Let $K(x,t_1,t_2)$, $x\in R^1$, $t_1<t_2$, denote the local time of the process B_1 at the point x in the time interval $[t_1,t_2]$. The one-dimensional field

$$Z_n=\int K(x,n,n+1)B_2(dx),\quad n=\dots,-1,0,1,\dots ,$$

where the integral in the last formula is an Itô integral, is a stationary self-similar field with self-similarity parameter $3/4$.

To see the self-similarity property one has to observe that

$$K(\lambda^{1/2}x,\lambda t_1,\lambda t_2) \overset{\Delta}{=} \lambda^{1/2}K(x,t_1,t_2) \quad \text{for all} \quad x\in R^1,\ t_1<t_2,\ \lambda>0$$

because of the relation $\lambda^{1/2}B_1(\lambda u) \overset{\Delta}{=} B_1(u)$. Hence

$$\sum_{j=0}^{n-1} Z_j \overset{\Delta}{=} n^{1/2} \int K(n^{-1/2}x,0,1)B_2(dx) \overset{\Delta}{=}$$

$$\stackrel{\Delta}{=} n^{3/4} \int K(x,0,1) B_2(dx) = n^{3/4} \, Z_0 \; .$$

The invariance of the multi-dimensional distributions of the field Z_n under the transformation (1.1) can be seen similarly.

To see the stationarity of the field Z_n we need the following two observations

a) $K(x,s,t) \stackrel{\Delta}{=} K(x+\eta(s),0,t-s)$ with $\eta(s)=-B_1(-s)$

(the form of η is not important for us. What we need is that the pair (η,K) is independent of B_2.)

b) If $\alpha(x), -\infty<x<\infty$, is a process independent of B_2 then

$$\int \alpha(x+u) B_2(dx) \stackrel{\Delta}{=} \int \alpha(x) B_2(dx) \quad \text{for all} \quad u \in R^1 \; .$$

It is enough to show, because of property a), that

$$\int K(x+\eta(s),0,t-s) B_2(dx) \stackrel{\Delta}{=} \int K(x,0,t-s) B_2(dx) \; .$$

This relation follows from property b), because the conditional distributions of the left and the right hand sides agree under the condition $\eta(s)=u$, $u \in R^1$.

The generalized field version of the above field Z_n is the field

$$Z(\varphi) = -\int [\int K(x,0,t) \frac{d\varphi}{dt} dt] B_2(dx), \quad \varphi \in S \; .$$

To explain the analogy of the fields Z_n and $Z(\varphi)$ we remark that the kernel of the integral defining Z_n can be written, at least formally, as

$$K(x,n,n+1) = \int \chi_{[n,n+1)}(u)\frac{d}{du} K(x,n,u)du ,$$

although K is a non-differentiable function. Substituting the function $\chi_{[n,n+1)}$ by $\varphi \in S$, and integrating by parts (or precisely, considering $\frac{d}{du}K$ as the derivative of a distribution) we get the above definition of $Z(\varphi)$.

Using the same idea as before, a more general class of self-similar fields can be constructed. The integrand $K(x,n,n+1)$ can be substituted by the local time of any self-similar field with stationary increments which is independent of B_2. Naturally, it must be clarified first that this local time really exists. One could enlarge this class also by integrating with respect to a self-similar field with stationary increments, independent of B_1. The integral with respect to a field independent of the field $K(x,s,t)$ can be defined without any difficulties.

There seems to be no natural way to represent the above fields as fields subordinated to a Gaussian field. On the other hand, the local times $K(x,s,t)$ are measurable with respect to B_1 , they have finite second moments, therefore they can be expressed by means of multiple Wiener-Itô integrals with respect to a white noise field. Then the process Z_n itself can also be represented via multiple Wiener-Itô integrals. It would be interesting to know whether the above defined self-

-similar fields, and probably a larger class of self-similar fields, can be constructed in a simple natural way via multiple Wiener-Itô integrals with the help of a randomization.

Section 7.

The definition of Wiener-Itô integrals together with the proof of Theorem 7.1 and Proposition 7.3 are given by Itô [17]. Theorem 7.2 is proved in Taqqu's paper [35]. He needed this result to show that the self-similar fields defined in [8] by means of Wiener-Itô integrals coincide with the self-similar fields defined in [35] by means of modified Wiener-Itô integrals.

Section 8.

The results of this section, with the exception of Theorem 8.6, are proved in [8]. Theorem 8.6 is proved in [21]. This paper was strongly motivated by [27]. Lemma 8.3 is formulated in a slightly more general form than Lemma 3 in [8]. The present formulation is more complicated, but it is more useful in some applications. We explain this in more detail. The difference between the original and the present formulation of this lemma is that here we allow that the integrand K_0 in the limiting stochastic integral is discontinuous on a small subset of $R^{k\nu}$, and the functions K_N may not converge on this set. This freedom can be exploited in some applications. Indeed, let us consider e.g. the self-similar fields constructed in Remark 6.5. In case $p<0$ the integrand in the formula expressing these fields is not continuous on the hyperplane $x_1+\dots+x_n=0$.

Hence, if we want to prove limit theorems where these fields appear as limit, and this happens e.g. in Theorem 8.6, then we can apply Lemma 8.3 but not its original version Lemma 3 in [8].

The example for non-central limit theorems given by Rosenblatt in [26] and its generalization by Taqqu in [32] are special cases of Theorem 8.2. In these papers only the special case $H_k(x)=H_2(x)=x^2-1$ is considered. Later Taqqu [35] prooved a result similar to Theorem 8.2', but he needed more restrictive conditions. The observation that Theorem 8.2' can be deduced from Theorem 8.2 is from Taqqu [32].

The method of [26] and [32] does not apply for the proof of Theorem 8.2 in the case of $H_k(x)$, $k\geq 3$. In these papers it is proved that the moments of the random variables Z_n^N converge the corresponding moments of Z_n^* . (Actually a different but equivalent statement is established.) This convergence of the moments implies the convergence $Z_n^N \xrightarrow{\mathcal{D}} Z_n^*$ if and only if the distribution of Z_n^* is uniquely determined by its moments.

Theorem 6.6 implies that the n-th moment of a k-fold Wiener-Itô integral equals to $\exp[\frac{k}{2}n \log n+O(n)]$. Hence some results about the so-called moment problem show that the distribution of a k-fold Wiener-Itô integral is determined by its moments only for $k=1$ and $k=2$. Therefore the method of moments does not work in the proof of Theorem 8.2 for $H_k(x)$, $k\geq 3$.

Throughout section 8 we have assumed that the correlation function of the underlying Gaussian field to which our fields are subordinated satisfies formula (8.1). This assumption seems natural for us, since it implies that the spectral measure of the Gaussian field satisfies Lemma 8.1, and such a condition is needed when $Z_{G_N}(dx)$ is substituted by $Z_{G_0}(dx)$ in the limit. It can be asked whether in Theorem 8.2 formula 8.1 can be substituted by the weaker assumption that the spectral measure of the Gaussian field satisfies Lemma 8.1. This question was investigated in section 4 of [8]. The investigation of the moments of Z_n^N shows that the answer is negative. The reason is that the validity of Lemma 8.1,unlike that of Theorem 8.2, does not depend on whether the spectral measure G has large singularities outside the origin or not. The discussion in [8] also shows that the Gaussian case, that is the case $H_k(x)=H_1(x)=x$ in Theorem 8.2, is considerably different from the non-Gaussian case. A forthcoming paper of M. Rosenblatt [28] gives a better insight into the above question.

The limiting fields appearing in Theorems 8.2 and 8.5 belong to a special subclass of the self-similar fields defined in Theorem 6.2. These results indicate that the self-similar fields defined in formula (6.5) have a much greater range of attraction if the homogeneous function f_n

in (6.5) is the constant function. The reason for the particular behaviour of these fields is that the constant function is analytic while a general homogeneous function typically has a singularity at the origin. A more detailed discussion about this problem can be found in [21].

References

[1] Bleher,P.M. and Missarov,M.D.(1980) Solution of Wilson's equation, and analytical renormalization. (in Russian) preprint of the Keldysh Institut of applied mathematics

[2] Bramson,M. and Griffeath,D.(1979) Renormalizing the three-dimensional voter model. Annals of Probability 7, 418-432.

[3] Cameron,R.H. and Martin,W.T.(1947) The orthogonal development of nonlinear functionals in series of Fourier-Hermite functionals. Ann.Math. 48, 385-392.

[4] Cramer,H.(1940) On the theory of stationary random processes. Ann.Math. 41, 215-230.

[5] Dawson,D. and Ivanoff, G.(1979) Branching diffusions and random measures. in Advances in Probability; Dekker, New York

[6] Dobrushin,R.L.(1979) Gaussian and their subordinated generalized fields. Annals of Probability 7, 1-28.

[7] Dobrushin,R.L.(1980) Automodel generalized random fields and their renorm-group. in Multicomponent random systems; Dekker, New York

[8] Dobrushin,R.L. and Major,P.(1979) Non-central limit theorems for non-linear functionals of Gaussian fields. Z Wahrscheinlichkeitstheorie verw.Gebiete 50, 27-52

[9] Dobrushin,R.L. and Major,P. On the limit behaviour of some self-similar fields. (in preparation)

[10] Dobrushin R.L.,Major,P. and Takahashi,J. Self-similar Gaussian fields.(in preparation)

[11] Dobrushin,R.L. and Minlos,R.A.(1977) Polynomials of linear random functions. Uspehi Mat.Nauk 32. 67-122.

[12] Dynkin, E.B.(1961) Die Grundlagen der Theorie der Markoffschen Prozesse. Springer Verlag Berlin-Göttingen-Heidelberg Band 108.

[13] Eidlin V.L. and Linnik,Ju.V.(1968) A remark on analytic transformation of normal vectors.Theor.Probability Appl. 13, 751-754 (in Russian)

[14] Gelfand,I.M. and Vilenkin,N.Ja.(1964) Generalized functions IV. Some application of harmonic analysis.

[15] Gross,L.(1975) Logarithmic Sobolev inequalities.Am.J. Math.97, 1061-1083.

[16] Holley,R.A. and Stroock,D.(1978) Invariance principles for some infinite particle systems in Stochastic Analysis. Academic Press, New York and London 153-173.

[17] Itô,K.(1951) Multiple Wiener Integral. J.Math.Soc. Japan 3, 157-164.

[18] Kesten,H. and Spitzer,F.(1979) A limit theorem related to a new class of self-similar processes. Z.Wahrscheinlichkeitstheorie und verw.Gebiete 50, 5-25.

[19] Kolmogorov,A.N.(1940) Wienersche Spirale und einige andere interessante Kurven im Hilbertschen Raum.C.R. (Doklady) Acad.Sci. U.R.S.S.(N.S.) 26 115-118.

[20] Löwenstein,J.H. and Zimmerman,W.(1975) The power counting theorem for Feynman integrals with massless propagators. Comm.Math.Phys.44, 73-86.

[21] Major,P. Limit theorems for non-linear functionals of Gaussian sequences. Submitted to Zeitschrift für Wahrscheinlichkeitstheorie und verw. Gebiete

[22] Major,P. Renormalizing the voter model. Space and space-time renormalization. Submitted to Studia Scientiarum Math. Hungarica.

[23] Mc.Kean,H.P.(1973) Geometry of differential space. Ann. Probab.1, 197-206.

[24] Mc.Kean,H.P.(1973) Wiener's theory of non-linear noise. in Stochastic Differential Equations SIAM-AMS Proc.6, 191-209.

[25] Nelson,E.(1973) The free Markov field. J.Functional Analysis 12, 211-227.

[26] Rosenblatt,M.(1962) Independence and dependence. Proc. Fourth Berkeley Symp.Math.Statist.Prob. 431-443 Univ. of California Press

[27] Rosenblatt,M.(1979) Some limit theorems for partial sums of quadratic forms in stationary Gaussian variables. Z.Wahrscheinlichkeitstheorie verw.Gebiete 49, 125-132.

[28] Rosenblatt,M. Limit theorems for non-linear functionals of Gaussian sequences. submitted to Z.Wahrscheinlichkeitstheorie verw. Gebiete.

[29] Segal,J.E.(1956) Tensor algebras over Hilbert spaces. Trans.Amer.Math.Soc.81, 106-134.

[30] Sinai,Ja.G.(1976) Automodel probability distributions. Theor.Probability Appl. 21, 273-320.

[31] Sinai,Ja.G.(198?) Mathematical problems of the theory of phase transitions. Akadémiai Kiadó, Budapest

[32] Taqqu,M.S.(1975) Weak convergence ot franctional Brownian Motion and to the Rosenblatt process. Z. Wahrscheinlichkeits theorie verw.Gebiete 31, 287-302.

[33] Taqqu,M.S.((1977) Law of the iterated logarithm for sums of non-linear functions of Gaussian variables. Z. Wahrscheinlichkeitstheorie verw.Gebiete 40, 203-238.

[34] Taqqu,M.S.(1978) A representation for self-similar process. Stochastic Processes Appl.7, 55-64.

[35] Taqqu,M.S.(1979) Covergence of iterated process of arbitrary Hermite rank. Z.Wahrscheinlichkeitstheorie verw.Gebiete 50, 27-52.

[36] Totoki,H.(1969) Ergodic Theory. Aarhus University Lecture Note Series 14.

Subject index

Notations

Vol. 700: Module Theory, Proceedings, 1977. Edited by C. Faith and S. Wiegand. X, 239 pages. 1979.

Vol. 701: Functional Analysis Methods in Numerical Analysis, Proceedings, 1977. Edited by M. Zuhair Nashed. VII, 333 pages. 1979.

Vol. 702: Yuri N. Bibikov, Local Theory of Nonlinear Analytic Ordinary Differential Equations. IX, 147 pages. 1979.

Vol. 703: Equadiff IV, Proceedings, 1977. Edited by J. Fábera. XIX, 441 pages. 1979.

Vol. 704: Computing Methods in Applied Sciences and Engineering, 1977, I. Proceedings, 1977. Edited by R. Glowinski and J. L. Lions. VI, 391 pages. 1979.

Vol. 705: O. Forster und K. Knorr, Konstruktion verseller Familien kompakter komplexer Räume. VII, 141 Seiten. 1979.

Vol. 706: Probability Measures on Groups, Proceedings, 1978. Edited by H. Heyer. XIII, 348 pages. 1979.

Vol. 707: R. Zielke, Discontinuous Čebyšev Systems. VI, 111 pages. 1979.

Vol. 708: J. P. Jouanolou, Equations de Pfaff algébriques. V, 255 pages. 1979.

Vol. 709: Probability in Banach Spaces II. Proceedings, 1978. Edited by A. Beck. V, 205 pages. 1979.

Vol. 710: Séminaire Bourbaki vol. 1977/78, Exposés 507–524. IV, 328 pages. 1979.

Vol. 711: Asymptotic Analysis. Edited by F. Verhulst. V, 240 pages. 1979.

Vol. 712: Equations Différentielles et Systèmes de Pfaff dans le Champ Complexe. Edité par R. Gérard et J.-P. Ramis. V, 364 pages. 1979.

Vol. 713: Séminaire de Théorie du Potentiel, Paris No. 4. Edité par F. Hirsch et G. Mokobodzki. VII, 281 pages. 1979.

Vol. 714: J. Jacod, Calcul Stochastique et Problèmes de Martingales. X, 539 pages. 1979.

Vol. 715: Inder Bir S. Passi, Group Rings and Their Augmentation Ideals. VI, 137 pages. 1979.

Vol. 716: M. A. Scheunert, The Theory of Lie Superalgebras. X, 271 pages. 1979.

Vol. 717: Grosser, Bidualräume und Vervollständigungen von Banachmoduln. III, 209 pages. 1979.

Vol. 718: J. Ferrante and C. W. Rackoff, The Computational Complexity of Logical Theories. X, 243 pages. 1979.

Vol. 719: Categorial Topology, Proceedings, 1978. Edited by H. Herrlich and G. Preuß. XII, 420 pages. 1979.

Vol. 720: E. Dubinsky, The Structure of Nuclear Fréchet Spaces. V, 187 pages. 1979.

Vol. 721: Séminaire de Probabilités XIII. Proceedings, Strasbourg, 1977/78. Edité par C. Dellacherie, P. A. Meyer et M. Weil. VII, 647 pages. 1979.

Vol. 722: Topology of Low-Dimensional Manifolds. Proceedings, 1977. Edited by R. Fenn. VI, 154 pages. 1979.

Vol. 723: W. Brandal, Commutative Rings whose Finitely Generated Modules Decompose. II, 116 pages. 1979.

Vol. 724: D. Griffeath, Additive and Cancellative Interacting Particle Systems. V, 108 pages. 1979.

Vol. 725: Algèbres d'Opérateurs. Proceedings, 1978. Edité par P. de la Harpe. VII, 309 pages. 1979.

Vol. 726: Y.-C. Wong, Schwartz Spaces, Nuclear Spaces and Tensor Products. VI, 418 pages. 1979.

Vol. 727: Y. Saito, Spectral Representations for Schrödinger Operators With Long-Range Potentials. V, 149 pages. 1979.

Vol. 728: Non-Commutative Harmonic Analysis. Proceedings, 1978. Edited by J. Carmona and M. Vergne. V, 244 pages. 1979.

Vol. 729: Ergodic Theory. Proceedings, 1978. Edited by M. Denker and K. Jacobs. XII, 209 pages. 1979.

Vol. 730: Functional Differential Equations and Approximation of Fixed Points. Proceedings, 1978. Edited by H.-O. Peitgen and H.-O. Walther. XV, 503 pages. 1979.

Vol. 731: Y. Nakagami and M. Takesaki, Duality for Crossed Products of von Neumann Algebras. IX, 139 pages. 1979.

Vol. 732: Algebraic Geometry. Proceedings, 1978. Edited by K. Lønsted. IV, 658 pages. 1979.

Vol. 733: F. Bloom, Modern Differential Geometric Techniques in the Theory of Continuous Distributions of Dislocations. XII, 206 pages. 1979.

Vol. 734: Ring Theory, Waterloo, 1978. Proceedings, 1978. Edited by D. Handelman and J. Lawrence. XI, 352 pages. 1979.

Vol. 735: B. Aupetit, Propriétés Spectrales des Algèbres de Banach. XII, 192 pages. 1979.

Vol. 736: E. Behrends, M-Structure and the Banach-Stone Theorem. X, 217 pages. 1979.

Vol. 737: Volterra Equations. Proceedings 1978. Edited by S.-O. Londen and O. J. Staffans. VIII, 314 pages. 1979.

Vol. 738: P. E. Conner, Differentiable Periodic Maps. 2nd edition, IV, 181 pages. 1979.

Vol. 739: Analyse Harmonique sur les Groupes de Lie II. Proceedings, 1976–78. Edited by P. Eymard et al. VI, 646 pages. 1979.

Vol. 740: Séminaire d'Algèbre Paul Dubreil. Proceedings, 1977–78. Edited by M.-P. Malliavin. V, 456 pages. 1979.

Vol. 741: Algebraic Topology, Waterloo 1978. Proceedings. Edited by P. Hoffman and V. Snaith. XI, 655 pages. 1979.

Vol. 742: K. Clancey, Seminormal Operators. VII, 125 pages. 1979.

Vol. 743: Romanian-Finnish Seminar on Complex Analysis. Proceedings, 1976. Edited by C. Andreian Cazacu et al. XVI, 713 pages. 1979.

Vol. 744: I. Reiner and K. W. Roggenkamp, Integral Representations. VIII, 275 pages. 1979.

Vol. 745: D. K. Haley, Equational Compactness in Rings. III, 167 pages. 1979.

Vol. 746: P. Hoffman, τ-Rings and Wreath Product Representations. V, 148 pages. 1979.

Vol. 747: Complex Analysis, Joensuu 1978. Proceedings, 1978. Edited by I. Laine, O. Lehto and T. Sorvali. XV, 450 pages. 1979.

Vol. 748: Combinatorial Mathematics VI. Proceedings, 1978. Edited by A. F. Horadam and W. D. Wallis. IX, 206 pages. 1979.

Vol. 749: V. Girault and P.-A. Raviart, Finite Element Approximation of the Navier-Stokes Equations. VII, 200 pages. 1979.

Vol. 750: J. C. Jantzen, Moduln mit einem höchsten Gewicht. III, 195 Seiten. 1979.

Vol. 751: Number Theory, Carbondale 1979. Proceedings. Edited by M. B. Nathanson. V, 342 pages. 1979.

Vol. 752: M. Barr, *-Autonomous Categories. VI, 140 pages. 1979.

Vol. 753: Applications of Sheaves. Proceedings, 1977. Edited by M. Fourman, C. Mulvey and D. Scott. XIV, 779 pages. 1979.

Vol. 754: O. A. Laudal, Formal Moduli of Algebraic Structures. III, 161 pages. 1979.

Vol. 755: Global Analysis. Proceedings, 1978. Edited by M. Grmela and J. E. Marsden. VII, 377 pages. 1979.

Vol. 756: H. O. Cordes, Elliptic Pseudo-Differential Operators – An Abstract Theory. IX, 331 pages. 1979.

Vol. 757: Smoothing Techniques for Curve Estimation. Proceedings, 1979. Edited by Th. Gasser and M. Rosenblatt. V, 245 pages. 1979.

Vol. 758: C. Năstăsescu and F. Van Oystaeyen; Graded and Filtered Rings and Modules. X, 148 pages. 1979.

Vol. 759: R. L. Epstein, Degrees of Unsolvability: Structure and Theory. XIV, 216 pages. 1979.

Vol. 760: H.-O. Georgii, Canonical Gibbs Measures. VIII, 190 pages. 1979.

Vol. 761: K. Johannson, Homotopy Equivalences of 3-Manifolds with Boundaries. 2, 303 pages. 1979.

Vol. 762: D. H. Sattinger, Group Theoretic Methods in Bifurcation Theory. V, 241 pages. 1979.

Vol. 763: Algebraic Topology, Aarhus 1978. Proceedings, 1978. Edited by J. L. Dupont and H. Madsen. VI, 695 pages. 1979.

Vol. 764: B. Srinivasan, Representations of Finite Chevalley Groups. XI, 177 pages. 1979.

Vol. 765: Padé Approximation and its Applications. Proceedings, 1979. Edited by L. Wuytack. VI, 392 pages. 1979.

Vol. 766: T. tom Dieck, Transformation Groups and Representation Theory. VIII, 309 pages. 1979.

Vol. 767: M. Namba, Families of Meromorphic Functions on Compact Riemann Surfaces. XII, 284 pages. 1979.

Vol. 768: R. S. Doran and J. Wichmann, Approximate Identities and Factorization in Banach Modules. X, 305 pages. 1979.

Vol. 769: J. Flum, M. Ziegler, Topological Model Theory. X, 151 pages. 1980.

Vol. 770: Séminaire Bourbaki vol. 1978/79 Exposés 525–542. IV, 341 pages. 1980.

Vol. 771: Approximation Methods for Navier-Stokes Problems. Proceedings, 1979. Edited by R. Rautmann. XVI, 581 pages. 1980.

Vol. 772: J. P. Levine, Algebraic Structure of Knot Modules. XI, 104 pages. 1980.

Vol. 773: Numerical Analysis. Proceedings, 1979. Edited by G. A. Watson. X, 184 pages. 1980.

Vol. 774: R. Azencott, Y. Guivarc'h, R. F. Gundy, Ecole d'Eté de Probabilités de Saint-Flour VIII-1978. Edited by P. L. Hennequin. XIII, 334 pages. 1980.

Vol. 775: Geometric Methods in Mathematical Physics. Proceedings, 1979. Edited by G. Kaiser and J. E. Marsden. VII, 257 pages. 1980.

Vol. 776: B. Gross, Arithmetic on Elliptic Curves with Complex Multiplication. V, 95 pages. 1980.

Vol. 777: Séminaire sur les Singularités des Surfaces. Proceedings, 1976-1977. Edited by M. Demazure, H. Pinkham and B. Teissier. IX, 339 pages. 1980.

Vol. 778: SK1 von Schiefkörpern. Proceedings, 1976. Edited by P. Draxl and M. Kneser. II, 124 pages. 1980.

Vol. 779: Euclidean Harmonic Analysis. Proceedings, 1979. Edited by J. J. Benedetto. III, 177 pages. 1980.

Vol. 780: L. Schwartz, Semi-Martingales sur des Variétés, et Martingales Conformes sur des Variétés Analytiques Complexes. XV, 132 pages. 1980.

Vol. 781: Harmonic Analysis Iraklion 1978. Proceedings 1978. Edited by N. Petridis, S. K. Pichorides and N. Varopoulos. V, 213 pages. 1980.

Vol. 782: Bifurcation and Nonlinear Eigenvalue Problems. Proceedings, 1978. Edited by C. Bardos, J. M. Lasry and M. Schatzman. VIII, 296 pages. 1980.

Vol. 783: A. Dinghas, Wertverteilung meromorpher Funktionen in ein- und mehrfach zusammenhängenden Gebieten. Edited by R. Nevanlinna and C. Andreian Cazacu. XIII, 145 pages. 1980.

Vol. 784: Séminaire de Probabilités XIV. Proceedings, 1978/79. Edited by J. Azéma and M. Yor. VIII, 546 pages. 1980.

Vol. 785: W. M. Schmidt, Diophantine Approximation. X, 299 pages. 1980.

Vol. 786: I. J. Maddox, Infinite Matrices of Operators. V, 122 pages. 1980.

Vol. 787: Potential Theory, Copenhagen 1979. Proceedings, 1979. Edited by C. Berg, G. Forst and B. Fuglede. VIII, 319 pages. 1980.

Vol. 788: Topology Symposium, Siegen 1979. Proceedings, 1979. Edited by U. Koschorke and W. D. Neumann. VIII, 495 pages. 1980.

Vol. 789: J. E. Humphreys, Arithmetic Groups. VII, 158 pages. 1980.

Vol. 790: W. Dicks, Groups, Trees and Projective Modules. IX, 127 pages. 1980.

Vol. 791: K. W. Bauer and S. Ruscheweyh, Differential Operators for Partial Differential Equations and Function Theoretic Applications. V, 258 pages. 1980.

Vol. 792: Geometry and Differential Geometry. Proceedings, 1979. Edited by R. Artzy and I. Vaisman. VI, 443 pages. 1980.

Vol. 793: J. Renault, A Groupoid Approach to C*-Algebras. III, 160 pages. 1980.

Vol. 794: Measure Theory, Oberwolfach 1979. Proceedings 1979. Edited by D. Kölzow. XV, 573 pages. 1980.

Vol. 795: Séminaire d'Algèbre Paul Dubreil et Marie-Paule Malliavin. Proceedings 1979. Edited by M. P. Malliavin. V, 433 pages. 1980.

Vol. 796: C. Constantinescu, Duality in Measure Theory. IV, 197 pages. 1980.

Vol. 797: S. Mäki, The Determination of Units in Real Cyclic Sextic Fields. III, 198 pages. 1980.

Vol. 798: Analytic Functions, Kozubnik 1979. Proceedings. Edited by J. Ławrynowicz. X, 476 pages. 1980.

Vol. 799: Functional Differential Equations and Bifurcation. Proceedings 1979. Edited by A. F. Izé. XXII, 409 pages. 1980.

Vol. 800: M.-F. Vignéras, Arithmétique des Algèbres de Quaternions. VII, 169 pages. 1980.

Vol. 801: K. Floret, Weakly Compact Sets. VII, 123 pages. 1980.

Vol. 802: J. Bair, R. Fourneau, Etude Géometrique des Espaces Vectoriels II. VII, 283 pages. 1980.

Vol. 803: F.-Y. Maeda, Dirichlet Integrals on Harmonic Spaces. X, 180 pages. 1980.

Vol. 804: M. Matsuda, First Order Algebraic Differential Equations. VII, 111 pages. 1980.

Vol. 805: O. Kowalski, Generalized Symmetric Spaces. XII, 187 pages. 1980.

Vol. 806: Burnside Groups. Proceedings, 1977. Edited by J. L. Mennicke. V, 274 pages. 1980.

Vol. 807: Fonctions de Plusieurs Variables Complexes IV. Proceedings, 1979. Edited by F. Norguet. IX, 198 pages. 1980.

Vol. 808: G. Maury et J. Raynaud, Ordres Maximaux au Sens de K. Asano. VIII, 192 pages. 1980.

Vol. 809: I. Gumowski and Ch. Mira, Recurences and Discrete Dynamic Systems. VI, 272 pages. 1980.

Vol. 810: Geometrical Approaches to Differential Equations. Proceedings 1979. Edited by R. Martini. VII, 339 pages. 1980.

Vol. 811: D. Normann, Recursion on the Countable Functionals. VIII, 191 pages. 1980.

Vol. 812: Y. Namikawa, Toroidal Compactification of Siegel Spaces. VIII, 162 pages. 1980.

Vol. 813: A. Campillo, Algebroid Curves in Positive Characteristic. V, 168 pages. 1980.

Vol. 814: Séminaire de Théorie du Potentiel, Paris, No. 5. Proceedings. Edited by F. Hirsch et G. Mokobodzki. IV, 239 pages. 1980.

Vol. 815: P. J. Slodowy, Simple Singularities and Simple Algebraic Groups. XI, 175 pages. 1980.

Vol. 816: L. Stoica, Local Operators and Markov Processes. VIII, 104 pages. 1980.